# Modular Mathematics

# Mechanics

# Module M1

by

Professor W.E. Williams

# Foreword

Professor Williams is the Chief Examiner for the Mathematics A2 paper and the Applied Mathematics modular papers for the Welsh Joint Education Committee and is also the Chief Examiner for a number of other Boards.

This book covers the M1 syllabus of the Welsh Joint Education Committee and only assumes the Pure Mathematical knowledge covered in the P1 syllabus.

An attempt has been made to present the material in such a way that students can make considerable headway before needing some of the slightly more sophisticated Pure Mathematical concepts.

In Chapters 2 and 3 only the elementary properties, associated with a right angled triangle, of the trigonometric functions have been assumed.

The material in Chapters 4 and 5, other than 4.4, 4.5 and 5.5 should be accessible without any knowledge of calculus. Similarly all the material in Chapter 7 (other than 7.1 where the formal definition of impulse is given) and the work on projectiles in Chapter 8 is not dependent on calculus.

This book could not have been written without the support, both moral and practical of my wife Janet. She read, and commented on in detail, the first draft and has carried out all the proof-reading. Thus every effort has been made to produce a text free from errors but if you should come across any that have escaped our attention it would be appreciated if you would kindly inform the publishers so that corrections may be incorporated in any subsequent edition.

# Contents

| | | Page No. |
|---|---|---|
| **Chapter 1** | Mechanics and Modelling | 1 |
| **Chapter 2** | Forces acting at a point | 7 |
| **Chapter 3** | Parallel forces acting on bodies | 56 |
| **Chapter 4** | Kinematics of Rectilinear Motion | 70 |
| **Chapter 5** | Dynamics of Rectilinear Motion | 96 |
| **Chapter 6** | Work, Energy and Power | 125 |
| **Chapter 7** | Impulse and momentum | 157 |
| **Chapter 8** | Motion under gravity in two dimensions | 181 |
| **Index** | | 204 |
| **Answers to Exercises** | | 206 |

# Chapter 1

# Mechanics and Modelling

After working through this chapter you should
- have some idea of the principles of modelling in Mechanics,
- be aware of some of the limitations of the particle model.

## 1.1 Basic principles of modelling

Mechanics is basically the study of what causes bodies to move and how they move. Statics deals with bodies at rest and Dynamics with bodies in motion. A knowledge of Mechanics is essential to design many of the things occurring in everyday living e.g. cars, aeroplanes, bridges, roads, houses. It is important, particularly when designing something new and complicated like a bridge or a car, to have some idea of whether the design will work. It is a bit late when the bridge is built to find a flaw!

The easiest way of finding whether the design is satisfactory is to try and formulate the real problem in mathematical terms and then use mathematical methods to predict what happens. This is what is known as constructing a mathematical model of a real situation and a mathematical model is effectively a simplified representation of a real world problem in mathematical terms. Real problems are often very complicated and in order to be able to do some useful mathematics it is usually necessary to make many simplifying assumptions. The idea is only to make simplifications which still retain the basic features of the problem. Once the assumptions have been made then mathematical calculations can be carried out and the predictions compared with any available experimental data.

If there is good agreement between theoretical predictions and experiment then the model has provided a good realisation of the real problem and a satisfactory solution has been obtained. This is the outer loop in the following diagram.

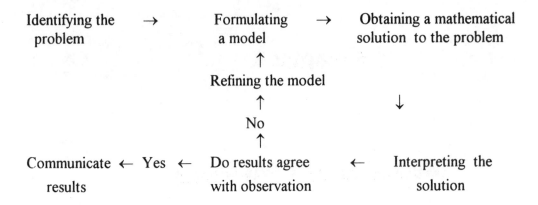

If the model does not agree with observation then the model has to be changed (refined) and the cycle repeated, possibly several times. This is the inner loop in the above diagram. The advantage of a mathematical model is that, once it has been shown to be valid for a range of parameters, it is possible to predict what happens, without further experiment, when parameters are changed. For example a mathematical model could be constructed for the motion of the bumper of a car when the car was involved in a crash. This model would involve a parameter describing the behaviour of the springing between the bumper and the car body. The model could be verified by crash tests for one, or more types of springing, and then the behaviour of the bumper for various types of springing determined, and the design incorporated in new cars without further tests to destruction.

In situations such as designing a new bridge or aeroplane then scale models would be built to compare the calculation with experiment. In many standard engineering applications however the modelling assumptions are generally well understood and have stood the test of time and in these cases the mathematical model can be used to produce particular design characteristics without the need for further tests.

In your Mechanics course you will have to model fairly simple situations by making standard asssumptions and you will have to be aware of the implications of these assumptions. For example you may be given the problem of a shot putter throwing the shot at a certain speed and asked to find where the shot hits the ground. You will learn that if the shot is modelled as a particle and the acceleration due to gravity is assumed to be constant then a fairly simple solution can be found. The path predicted by the model will be that shown by the left hand diagram, the actual path will be something like that in the right hand diagram.

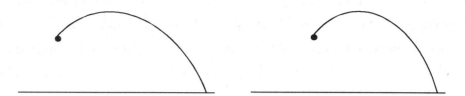

It may not be obvious which of the modelling assumptions should be changed in order to obtain agreement with observation. You are generally aware that, as you walk and run, you experience some air resistance and therefore it might be reasonable to refine the model to take this into account. This is not particularly easy and different types of air resistances have to be considered. It is possible to eventually arrive at a more accurate model but the calculations are complicated. You should however be aware that, when air resistance is taken into account, the predicted horizontal distance travelled by the shot is decreased. Calculations on the more complicated model show that under typical conditions the error in neglecting resistance is about 3%. In practice neither the speed of the throw, nor the angle of projection, would be known to this level of accuracy and therefore it is pointless to try and construct an elaborate model in these circumstances. It is important, when modelling, not to try and set up an elaborate model when the data to be used in the model is not particularly accurate.

There are circumstances where extreme accuracy is necessary and where the data is known accurately. An example of this is the free flight path of a space shuttle returning to earth. In this case it is necessary to assume, to get the required level of accuracy, the exact form of the acceleration of the earth, and also take into account corrections due to the rotation and the curvature of the earth.

In the Glossary in Chapter 2 most of the standard assumptions generally made are described and a more detailed account of the assumptions is given in 2.4.

There is one particular modelling assumption which is made throughout most of the book and this is modelling a body as a particle and the assumptions and limitations of this model will now be considered.

## 1.2 Implications of particle model

In practically all the problems that you will come across in your course and certainly in all problems involving motion all bodies will be modelled as particles. A particle is effectively something with no size but with mass or weight. You will be given the precise definitions of these in subsequent chapters but you will already have some rough idea of weight.

Effectively a particle is represented by a point. There are two basic assumptions implicit in using the particle model. The first assumption is an essentially geometric one that the dimensions of the body are small compared with the other dimensions involved in the problem. Therefore, for example, in tracking an aircraft it is reasonable to represent it by a point since knowing the position of a point of an aircraft would give you a very good idea of the position of the whole aircraft. This is roughly what happens on radar screens tracking aircraft and ships. In the context of space the aircraft is very much smaller than the distance from a tracking station. Similarly in analysing the motion of the earth round the sun it is sufficient, in view of the distances involved, to represent the earth as a particle.

It is very important to realise, when modelling, that the model that you need to choose on a particular occasion depends on the occasion and on what you are trying to find. For example a particle model of a car is sufficient to give information about its position and speed but it would be completely inadequate for road design. In the latter case it is necessary to calculate the overtaking sight distance (the distance open to view of a car travelling at the design speed of the road and wanting to overtake slower traffic without causing an obstruction) and so the model would have to take into account the lengths of vehicles.

The other assumption of the particle model of a body is that effectively the motion of one point of the body is completely representative of the motion of the whole body. This is not always true as you can see, for example, by throwing up a paper plate. All the points eventually move downwards but there will be a considerable amount of wobbling and certainly not all points of the plate move in the same way. It is possible to prove that the general motion of a body involves both a direct motion (translation) and a rotation. You can see this by throwing up a ball since in most circumstances there will also be a rotation or spin.

The particle model completely ignores the fact that not all points move in the same way and so is insufficient to model some situations. To a large extent the motion parallel to the road of all points of a car is the same and therefore the particle model is adequate. There will however be some slight motion perpendicular to the road, due to the effect of the suspension, and a model involving a box attached to four springs would be necessary to analyse this motion.

A simple example of the inadequacy of the particle model is when you have a tall piece of furniture on a rough floor and you push it near the top. The particle model predicts that the furniture slides as one piece but the reality is that it sometimes topples!

Another example is the motion of a snooker ball. The particle model would always predict that the ball would go in a straight line but in actual fact it can be made to go in a curve. A model taking into account spin predicts the curved motion. You may have seen golf balls struck into a hole travel round the inside of the hole and then come out, this would not be predicted by the particle model.

The saving grace of the particle model for problems involving motion is that there does exist for any body a point, whose motion is exactly that of a particle of mass equal to that of the body and acted on by all the external forces acting on the body. For most balls this point will be at its geometric centre. Therefore the particle model will give a very good estimate of where a ball lands but give very little information about what happens after because, as you have seen in ball games, the spin on the ball has an enormous effect on the path after impact.

## 1.3 Refinements of modelling

In order to refine a model you have to be very clear what your initial modelling assumptions are and then see which of them can, or need, to be changed. The most frequent modelling assumption is to ignore friction or air resistance and, as you will find later, it is often possible to refine, fairly easily, the model to take these into account. Even when you are not required to carry out detailed calculation you should be aware of the qualitative implications of the refinement. For example, if a ball is thrown vertically upwards, the maximum height it reaches will be less than that predicted by a model ignoring air resistance. The latter model therefore overestimates the maximum height reached. This point will be explained in more detail in Chapter 5.

The modelling assumptions used in constructing the table of stopping distances in the Highway Code assume that the road is flat. The model has to be refined to take into account whether there is an uphill or downhill slope. The particle model of a car shows that the maximum slope on which a car can be parked without slipping is independent of whether the front wheels are pointing down or up the hill. A more accurate model taking the car as a box on wheels shows that this is not always the case.

It is sometimes relatively easy to refine the particle model so as to take some account of the size of a body. For example you will learn later in the course how to find the time taken for a ball (modelled as a particle), struck from a floor, to a hit a vertical wall at right angles as shown in the left hand diagram below.

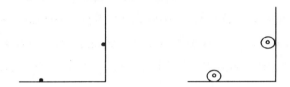

If the ball is of radius $a$ you can take the size into account by representing the ball by a particle starting at a point at a distance $a$ above the floor, as in the right hand diagram, and finding the time until it is at a horizontal distance $a$ from the wall.

It is however difficult to take account of the spin of a body in such a relatively simple way.

# Chapter 2

# Forces acting at a point

After working through this chapter you should
- have a clear idea of what a force is,
- understand what is meant by the resultant of several coplanar forces acting at a point and be able to find the resultant and its components in two perpendicular directions,
- appreciate some of the ideas used in modelling the effects of forces on small bodies,
- be able to solve problems relating to the equilibrium of a small body,
- be able to apply the laws of friction in simple cases.

## 2.1 Forces

Most people have an intuitive idea of a force as a "push" or a "pull". Taking this idea further, pushing or pulling, for example, a book on a desk will make the book move. Therefore, in this case, a force has caused a change in the motion of the book.

There are many other instances of what would be instinctively regarded as a force producing a change in motion, for example a tennis player hitting a ball or a driver braking a car.

The idea of a force as something which changes the motion of a body actually defines a force. On this definition most, but not all, forces conform to the idea of a "push" or a "pull". There are essentially two classes of forces:- contact forces where the change in motion is produced by direct contact, and non-contact forces where there is no direct physical link to the body. The most obvious example of the latter is the force due to gravity: this does however conform to the the simple idea of a "pull" since, if you jumped from a wall, you would certainly experience a pull!

## Forces acting at a point

Though a force is defined as "that which changes the motion of a body" it is not always true that applying a force will change the motion. For example pushing a brick wall will not bring it down.

This is because there are other forces binding the wall together and you cannot push sufficiently hard to overcome them. If a heavy lorry hits a wall then the chances are high that the wall would move. When you push a wall you will feel a resistance, i.e. the wall pushes back. This is an example of Newton's third law which states that "to every action there is an equal and opposite reaction". The tennis player also feels this reaction when she hits the ball.

Another example where you may think that there are forces acting but there is no change in motion is a car moving at a steady speed.

There is a driving force $F$ exerted by the wheels but there is also a drag force $D$ due to air resistance and when the speed is not changing the forces balance, i.e. the nett force is zero.

The exact relation between force and motion is provided by Newton's second law of motion (5.1), which states that the force acting on a moving particle is proportional to the particle's acceleration.

### Measuring force

One of the simplest ways of measuring some contact forces is by using a spring balance. This is a device where the top end of a spring is fixed, a hook is attached to the bottom end and the spring is allowed to hang freely. The diagram shows an old fashioned type of spring balance that you may still sometimes see and you probably have seen smaller balances in your science courses or Mechanics kits.

*Forces acting at a point*

If you pull down on the hook then the spring extends and a small needle attached to it moves down: if you pull harder then the needle drops further. The spring balance is a good method of measuring force since, for forces of reasonable size, the extension of the spring is directly proportional to the force. (This is Hooke's law which you will meet in 2.2). The balance can be calibrated by taking a particular force as the unit of force and the extensions corresponding to different forces are then marked out on the scale. The spring balance is of limited use in measuring forces in practical situations but it is important, for theoretical reasons, to know that a unit of force can be defined independently of motion.

The most commonly used unit of force is the newton abbreviated to N, so that a force of six newtons would be written as 6 N. The formal definition of the newton is given in (5.1). A thousand newtons is denoted by 1 kN.

**Representing forces**

If a small body is attached to a thin rod and the other end of the rod is pulled, then the body will move in the direction of the rod. The more effort exerted on the rod the more rapid the motion. If the rod is pulled in a different direction then the body will move in a different direction. If, in either case, the rod were pushed, not pulled, the motion would be in the opposite direction.

The rod, since it produces a change in motion, exerts a force on the body and this force has both magnitude (more effort produces faster motion) and direction (the different directions of the rod). Therefore a force can be represented by a directed line, with the length of the line representing the magnitude of the force and its direction, usually shown in diagrams by an arrow head, being that of the force. Since pushing and pulling produces different motions it is important to show the direction along the line in which the force is acting. The two forces shown, though acting along parallel lines and equal in magnitude, are different from each other since their directions are opposite.

*Forces acting at a point*

Quantities having both magnitude and direction are called vectors and you will learn more about them in book (M2).

In order to distinguish, in print, between a force and its magnitude bold type will be used to refer to the force and italic type for the magnitude. Therefore **F** refers to a force and $F$ to its magnitude. You could do this in writing by underlining when you refer to the force and not when you are referring to its magnitude. If a force **G** is equal in magnitude to another one **F** and acts along the same, or parallel, line but in the opposite direction then **G** could be denoted by -**F**. Therefore if the left hand force in the above diagram is **F** then the right hand force is -**F**.

**Combining forces**

If two rods were attached to a body and both rods were pulled then, as long as the pulls were not equal and opposite, the body would still move. The direction of the motion would not usually be parallel to either rod, but somewhere in between them. There is therefore a force acting on the body. This is called the resultant of the two forces and it is a single force that has the same effect as the two forces. Obviously more than two rods could be attached and :

**The resultant of any number of forces acting at a point is the single force which has the same effect (i.e produces the same motion) on a small body placed at that point.**

**Resultant of two forces acting at a point**

The rule for finding the resultant of two forces is a slightly unusual one and is as follows.

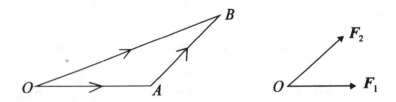

If there are two forces $F_1$ and $F_2$ acting at the point $O$ as shown then if the line $OA$ represents $F_1$ and the line $AB$ represents $F_2$ the line $OB$ represents the resultant of the two forces. This is, for obvious reasons, often called the "triangle rule" for combining (or adding) forces.

An alternative way of expressing this rule is to say that the resultant is represented, as shown above, by the diagonal of the parallelogram formed by the lines representing the two forces (this is the "parallelogram rule").

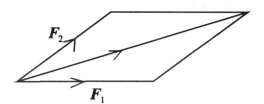

This rule can be shown to be a consequence of Newton's law of motion but it can also be verified experimentally. You may have seen a verification carried out in one of your science courses. A possible experimental set up is described in Miscellaneous Exercises 2, q 23.

For two parallel forces the rule simplifies, since $OB$ is then a straight line:
(a) for two forces acting in the same sense the resultant acts in the same sense and its magnitude is the sum of the magnitudes;
(b) for two forces acting in the opposite sense the resultant acts in the sense of the force with the greater magnitude and its magnitude is the positive difference of the magnitudes.
(c) for two forces of equal magnitude but of opposite direction the resultant force is zero. (You can check this rule by trying particular cases).

### Example 2.1
Find the resultant of the following systems of forces

$$3\,N \longrightarrow \quad 5\,N \qquad 5\,N \longleftarrow \longrightarrow 7\,N \qquad 9\,N \longleftarrow \longrightarrow 5\,N \qquad 4\,N \longleftarrow \longrightarrow 4\,N$$
$$\quad\text{(a)} \qquad\qquad\qquad \text{(b)} \qquad\qquad\qquad \text{(c)} \qquad\qquad\qquad \text{(d)}$$

(a) Both forces are acting in the same direction so the resultant acts to the right and is of magnitude 8 N.
(b) The forces are acting in the opposite direction, the force acting to the right has the greater magnitude and therefore the resultant acts to the right and is of magnitude 2 N.

(c) The forces are acting in the opposite direction, the force acting to the left has the greater magnitude and therefore the resultant acts to the left and is of magnitude 4 N.

(d) The forces are acting in opposite directions, both have the same magnitude and therefore the resultant is zero.

**Example 2. 2**

There are two perpendicular forces of magnitudes 3 N and 4 N acting at a point $O$. Find their resultant.

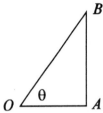

In the above diagram $OA$ represents the force of magnitude 3 N and $AB$ represents the force of magnitude 4 N. Their resultant will be represented by $OB$. By Pythagoras' theorem its magnitude will be $\sqrt{3^2 + 4^2}$ N = 5 N. The angle $\theta$ is given by

$$\tan \theta = \frac{4}{3}.$$

Using the $\tan^{-1}$ function on your calculator gives $\theta$ to be approximately 53.1°.

Example 2.2 shows that a force represented by $OB$ (5 N acting at an angle of approximately 53.1° to a line going across the page from left to right, is the resultant of two perpendicular forces of magnitudes 3 N and 4 N respectively. These two forces are referred to as perpendicular components of the force along $OB$.

**Components of a force**

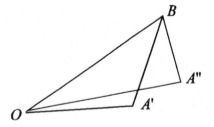 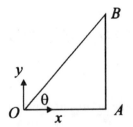

In the left hand diagram above $OB$ represents a given force $F$. An infinite number of triangles can be drawn from points $A'$, $A''$, $A'''$ etc onto $OB$ as base so that $F$ can be

*Forces acting at a point*

regarded as the resultant of forces represented by $OA'$ and $A'B$, or $OA''$ and $A''B$, etc. Therefore there are an infinite number of ways in which $F$ can be expressed in terms of two separate forces or components. The process of representing a force in terms of two components is referred to as resolving the force into its components (or into its resolved parts).

The most usual way of resolving a force is into components along two perpendicular lines which gives a right angled triangle such as $OAB$ in the right hand diagram above. For ease of reference the $x$ and $y$ axes of a coordinate system are chosen to be in the directions of $OA$ and $AB$ respectively. The component in the $x$-direction (called the $x$-component) of the force $F$ represented by $OB$ is denoted by $X$ and the component of the force in the $y$-direction (i.e. the $y$-component) by $Y$. The triangle rule for resultants states if the length of $OA$ is proportional to $X$ and the length of $AB$ is proportional to $Y$ then $OB$ is in the direction of $F$ and its length is proportional to $F$.

The angle between the $x$-direction and $F$ is denoted by $\theta$. The triangle $OAB$ is right angled and you know, from the definitions of sine and cosine, that
$\dfrac{OA}{OB} = \cos\theta$, and therefore $\dfrac{X}{F} = \cos\theta$, i.e. $X = F\cos\theta$.
Similarly $\dfrac{AB}{OB} = \sin\theta$ so that $\dfrac{Y}{F} = \sin\theta$, i.e. $Y = F\sin\theta$.

The component of a force $F$ in a given direction can now be defined as $F\cos\alpha$, where $\alpha$ is the angle between $F$ and the given direction. This is consistent with the above results since the angle between $F$ and the positive $x$-direction is $\theta$ and the angle between $F$ and the positive $y$-direction is $\dfrac{\pi}{2} - \theta$ giving the $y$-component as $F\cos(\dfrac{\pi}{2} - \theta) = F\sin\theta$, since $\cos(\dfrac{\pi}{2} - \theta) = \sin\theta$. This definition of component also shows that, when the angle between the force and the reference direction is obtuse, the component can be negative.

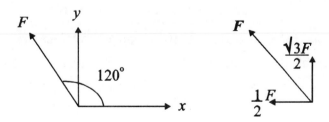

This occurs in the diagram above where $\theta = 120°$ so that $X = F\cos(120°) = -\frac{1}{2}F$ and $Y = F\cos(30°) = \frac{F\sqrt{3}}{2}$. All that this means is that the force is the resultant of $\frac{1}{2}F$ to the left and $\frac{F\sqrt{3}}{2}$ up the page.

You need to be very careful about the signs of the components. One way of doing this is by being very careful to pick the correct angle and take its cosine correctly. Another way is to resolve the force into two positive perpendicular components along the lines in which you are required to find components. This is shown in the diagram above. It is worth remembering that if $\theta$ is the acute angle with one of the lines then the positive components along the lines are $F\cos\theta$ and $F\sin\theta$. If the directions of these components are opposite to the actual direction in which you have to find the component then the component in the required direction is found by changing the sign. In the above diagram the component to the left is $\frac{1}{2}F$ and so the component to the right is $-\frac{1}{2}F$. When you start to find components it is a good idea to show both the force and its positive components as described above.

### Example 2.3
Find the x- and y- components of the following forces.

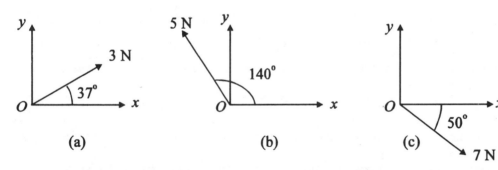

(a)                  (b)                  (c)

(a) The angles between the force and the x and y axes are 37° and 53° respectively so the components are $3\cos(37°)$ N $= 2.40$ N and $3\cos(53°)$ N $= 1.81$ N.

(b) The angles between the force and the positive x and y directions are 140° and 50° respectively so the components are $5\cos(140°)$ N $= -3.83$ N and $5\cos(50°)$ N $= 3.21$ N. Alternatively the positive components along the x- and y- axes are shown in the left hand diagram below.

*Forces acting at a point*

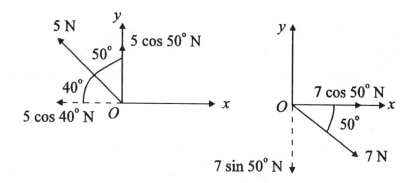

The components to the left and upwards are 5 cos (40°) = 3.83 N and 5 cos (50°) N {or 5 sin 40° N} = 3.21 N. Therefore the *x*- and *y*- components are = −3.83 N and 3.21 N.
(c) The angles between the force and the positive *x* and *y* directions are 50° and 220° respectively so the components are 7 cos (50°) N = 4.50 N and 7 cos (220°) N = −5.36 N. Alternatively the positive components along the *x*- and *y*- axes are shown in the right hand diagram above.

The components to the right and down are 7 cos (50°) N = 4.50 N and 7 sin (50°) N = 5.36 N. Therefore the *x*- and *y*- components are 4.50 N and −5.36 N.

### Exercises 2. 1

In all numerical questions answers should be given to three significant figures. Find the *x*- and *y*- components of the forces shown in questions 1 to 4.

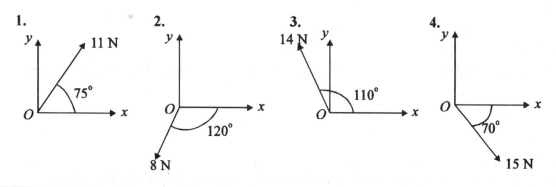

5.

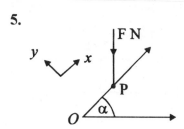

*Forces acting at a point*

The diagram represents a section of a plane inclined at an angle $\alpha$ to the horizontal and a force of magnitude $F$ N acts vertically downwards at a point $P$ of the plane. Find the $x$- and $y$- components of this force, referred to the axes shown which are parallel and perpendicular to the plane, when

(a) $F = 10$, $\alpha = 30°$, (b) $F = 6$, $\alpha = 50°$.

**6** Carry out the same calculations as in 5 when the vertical force is replaced by a horizontal one acting to the right and of magnitude $Q$ N when

(a) $Q = 8$, $\alpha = 60°$, (b) $Q = 4$, $\alpha = 70°$.

**Components of the resultant of forces acting at a point**

For several forces acting at a point, the component of their resultant in a particular direction is the sum of the components of the separate forces in that direction.

No attempt will be made to prove this result in general but you can see, from the following diagram, that it is certainly true for two forces in the simple case shown.

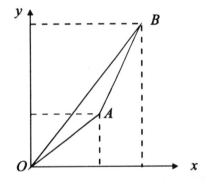

In Examples 2.4 to 2.6 the $x$-direction is taken to be to the right across the page and the $y$-direction to be up the page.

**Example 2.4**

Find the $x$- and $y$-components of the resultant of the following forces acting at a point.

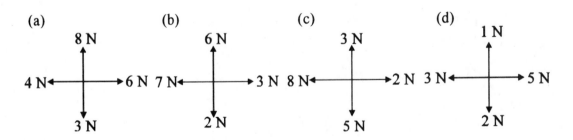

(a) The total *x*-component is 6 N–4 N = 2 N, whilst the *y*-component is 8 N–3 N = 5 N.
(b) The total *x*-component is 3 N–7 N = –4 N, whilst the *y*-component is 6 N–2 N = 4 N.
(c) The total *x*-component is 2 N–8 N = –6 N, whilst the *y*-component is 3 N–5 N = –2 N.
(d) The total *x*-component is 5 N–3 N = 2 N, whilst the *y*-component is 1 N–2 N = –1 N.

**Example 2.5**

Find the *x*- and *y*-components of the resultant of the forces of magnitude 3 N and 4 N, acting at the point *O*, as shown in the left hand diagram below.

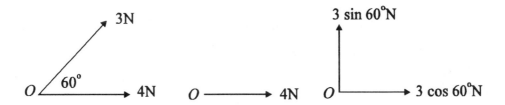

The right hand diagrams show each force resolved into its positive components across and along the page. The *x*- and *y*- components of the resultant therefore are $4 + 3 \cos 60°$ N = 5.5 N and $3 \sin 60°$ N = 2.60 N.

**Example 2.6**

Fnd the *x*- and *y*- components of the resultant of the following forces acting at a point *O*.

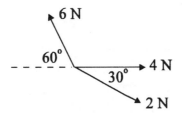

Each force can be resolved into components as shown below.

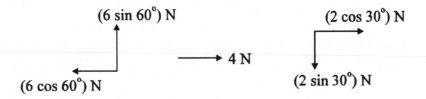

The force of magnitude 6 N has a negative component in the positive *x*- direction of $-6 \cos 60°$ N, whilst the force of magnitude 2 N has a negative component in the *y*-direction of $-2 \sin 30°$ N.

The x- and y- components of the resultant are therefore 2 cos 30° + 4 − 6 cos 60° N = 2.73 N and 6 sin 60° − 2 sin 30° N = 4.20 N.

**Exercises 2.2**

In each of the following problems find the x- and y- components of the resultant of the forces shown acting at a point $O$.

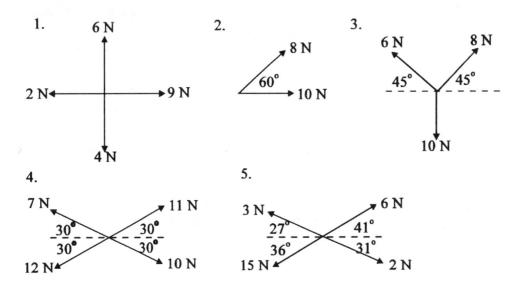

**Equilibrium problems**

When a number of forces acting at a point have zero resultant the forces are said to be in equilibrium. Since a point (particle) is a model for a small body you can say that, if the resultant force is zero, the point (particle, body) is in equilibrium. A body in equilibrium is at rest.

Forces acting on a particle will be in equilibrium if the sum of the components, in two non parallel directions, of all the forces acting is zero.

(This follows from the triangle rule since if a triangle has two sides zero its third side must also be zero.)

The condition that components in two directions are zero for equilibrium is the one that is generally easiest to apply in problems and it is used in Examples 2.7 to 2.9. For problems involving only three forces there is an alternative geometric method which is sketched after Example 2.9.

In order to have some practice with the general technique it is useful to look first at a few problems which are basically numerical. Practical problems requiring modelling and the use of given physical conditions are given in 2.2.

## Example 2.7

Find $R$ such that the forces shown below are in equilibrium.

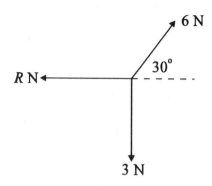

The force of magnitude 6 N has components $6\cos 30°$ N and $6\sin 30°$ N to the right and up the page respectively.

The unknown force has a component of magnitude $R$ N to the left and the third force has a component of magnitude 3 N down the page. Paying attention to the senses of the various forces, the components of the resultant to the right and up the page are $6\cos 30° - R$ N and $6\sin 30° - 3$ N = 0 N. The component up the page is already zero and, therefore, for equilibrium, the other one must also be zero so

$$R = 6\cos 30° \text{ N} = 3\sqrt{3} \text{ N}.$$

An alternative solution to this example is given after Example 2.9.

## Example 2.8

Find $P$ and $Q$ such that the system of forces shown in the left hand diagram below is in equilibrium.

The separate forces have components as shown in the right hand diagram. On taking account of the directions of the components, the $x$- and $y$- components of the resultant are $(P\cos 60° + 2\cos 30° + 4\cos 60° - Q)$ N and $(P\sin 60° + 2\sin 30° - 4\sin 60°)$ N respectively.

Both components have to be zero. Equating the second one to zero gives
$P \sin 60° = 4 \sin 60° - 2 \sin 30°$ so that $P = 2.85$. Equating the first component to zero gives $Q = P \cos 60° + 2 \cos 30° + 4 \cos 60°$ and, finally substituting for $P$, $Q = 5.16$.

As you get more familiar with using components you will not need to use separate diagrams to work out the components of the various forces and should be able to carry out the calculations in your head. It is also a matter of preference whether you work out the total component in a given direction and equate it to zero or, for example, equate the components to the left (or up) to those to the right (or down). The "shorthand" for saying what you are doing is resolving parallel and perpendicular to a given direction (you should always say clearly the direction in which you are resolving).

The choice of the directions in which to resolve is yours and you can sometimes make life easier by resolving perpendicular to an unknown force since this force will not have a component perpendicular to itself since $\cos 90° = 0$.

**Example 2.9**

Find the values of $P$ and $Q$ so that the system shown below is in equilibrium.

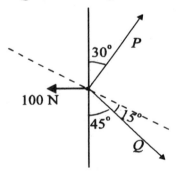

Resolving perpendicular to the force of magnitude 100 N gives
$$P \cos 30° = Q \cos 45°,$$
resolving parallel to the above force gives
$$P \cos 60° + Q \cos 45° = 100 \text{ N}.$$
These equations have now to be solved for $P$ and $Q$ and the final answers are
$$P = 73.2 \text{ N}, \quad Q = 89.7 \text{ N}.$$
This problem can actually be solved without getting two simultaneous equations by resolving along the perpendiculars to the unknown forces. The dashed line is perpendicular to the force of magnitude $P$ and the forces of magnitudes $Q$ and 100 N make angles of 15° and 30° with this line. Resolving parallel to the dashed line gives
$$Q \cos 15° = 100 \cos 30° \text{ N},$$
so that
$$Q = 89.7 \text{ N}.$$

Resolving along the perpendicular to the force of magnitude $Q$ gives

$$P \cos 15° = 100 \cos 45° \text{ N},$$

so that $\quad\quad\quad\quad\quad\quad P = 73.2 \text{ N}.$

This method avoids the algebra of solving two equations but requires care in getting the right angles and some of you might prefer the algebra to the calculation of angles!

**Triangle of forces**

For equilibrium problems where there is only one unkown force an alternative method is to find the resultant of the known forces. The unknown force is then of the same magnitude as this resultant but opposite in direction. This method is considered in more detail in 2.3 but, for three forces acting at a point, it reduces to a geometric one. The triangle rule gives that the resultant of two forces represented in magnitude and direction by $OA$ and $AB$ is represented by $OB$. Therefore the third force necessary for equilibrium is represented by $BO$. Therefore if $OA$ is drawn to represent one force, $AB$ to represent the second force and $BO$ to represent the third a closed triangle is formed as shown below.

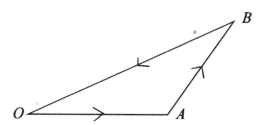

This will only be true if the three forces are in equilibrium, for any three arbitrary forces a closed triangle would not be formed. The triangle is known as the "triangle of forces" and in forming it you must draw the lines representing the forces "in order". This means that the arrows must follow each other and the arrow heads must not point towards each other. Several forces acting at a point form a closed polygon, the "polygon of forces" when they are in equilibrium but calculations for such polygons can be complicated. You are not yet able to solve problems involving calculating the sides and angles of a general triangle but you can use the triangle of forces when two forces are perpendicular to each other. (You can solve problems for triangles by scale drawing but you should not do this in a Mathematics examination unless you are specifically told that scale drawing is acceptable)

The method will be illustrated by re-doing Example 2.7.

## Example 2.7 (alternative solution)

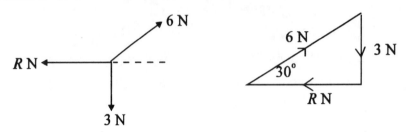

The forces acting are shown in the left hand diagram and the triangle of forces formed in the right hand one.

From the triangle of forces

$$\cos 30° = \frac{R}{6} \quad \text{i.e. } R = 6\cos 30° = 3\sqrt{3}.$$

## Exercises 2.3

Each of the problems 1 to 7 shows a system of forces in equilibrium. Find the unknowns in each case, giving forces to three significant figures and each angle to the nearest degree.

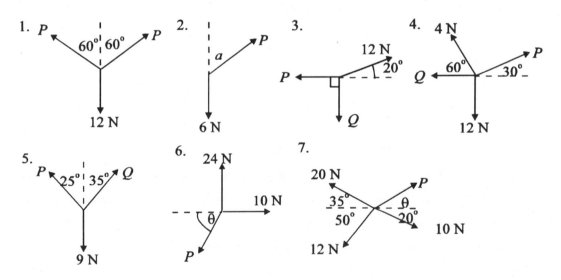

8 Three forces of magnitudes $X$, $Y$ and $R$ are shown in equilibrium at a point.

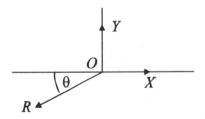

Find $R$ and $\tan \theta$ when
(i) $X = 4$ N, $Y = 3$ N
(ii) $X = 7$ N, $Y = 2$ N.

## 2.2 Equilibrium problems involving physical modelling

All the problems in 2.1 have been straightforward because the various forces acting were prescribed exactly for you. In practical situations this will not be the case. You will have to use the data to model the problem and decide what forces are acting and then end up with the type of problems that you have been solving.

### Modelling forces in the physical world

In setting up models of practical situations it is usually necessary to make simplifying assumptions about various objects and the forces exerted by, or on, them. These assumptions are often then summarised in a simple phrase or word (e.g particle, light string). It is very important that you realise precisely what assumptions are implied in using a particular description. The assumptions associated with the most common phrases used are summarised below in the form of a Glossary. This Glossary is intended to help you in interpreting questions that you have to answer. It should not be followed blindly. You should also read the more detailed explanation in 2.4 of the modelling assumptions that are summarised in the Glossary.

Some of the descriptions are a convenient fiction in that the situations described do not actually exist but, nervertheless, they often approximate to reality.

One major modelling assumption that you will be forced to make, since you have as yet only learnt methods for tackling problems where forces are acting at a point, is that any body that you consider has to be treated as a point. Therefore small objects like parcels and large ones like ships have to be modelled in the same way

At a first reading you may not want to go into the reasoning in 2.4 behind the assumptions in 2.4 but you should look at the Glossary where the assumptions corresponding to various phrases are summarised.

*Forces acting at a point*

**Glossary:**

**Force of gravity:** Normally assumed to be constant and acting in the downwards vertical direction, whose magnitude is the weight of the body.

The weight of a body of mass $m$ kg is $mg$ N where $g$ is the magnitude, approximately 9.81, of the acceleration of gravity in ms$^{-2}$. In most of the calculations in this book the approximation 9.8 will be used.

For a general body the force of gravity acts through the centre of gravity. The location of this is a geometrical property of the body and for a uniform straight rod it is at its midpoint.

A simple way of measuring the weight of a body is by using bathroom scales. The reading on the scales is actually the reaction of the scales on the body. When the scales are stationary this is equal to the weight of the body.

**Light** A light body is one with zero weight or, more realistically, one whose weight is negligible compared to the other forces acting.

**Strings** are represented by thin lines and used to model ropes and even chains. When taut they exert a force, the tension, inwards from their ends, on bodies attached at the ends.

The tension in a string is effectively the pulling force exerted by one part of a string on the other part and can vary over the length of the string.

If a string is passed round any body such as a peg (or pulley), as shown in the diagram, then

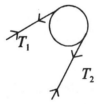

the forces exerted on the peg are the tensions $T_1$ and $T_2$, acting at the points of tangency as shown.

Strings cannot exert a push away from their ends, nor exert any force perpendicular to themselves. There is no tension in a slack string.

If your calculations result in a negative or zero tension then you will have made a mistake.

**Light strings** The tension is constant throughout the length of a light string and, if taut, the string will be straight.

**Inextensible strings,** this means that the length does not change when a force is applied to the ends but the more realistic view is that the length does not change sufficiently for the tension to change.

## Forces acting at a point

**Extensible (elastic) strings** The length of such strings can vary and the tension will depend on the length.

The normal modelling assumption is that the tension ($T$) is directly proportional to the extension ($x$) this is **Hooke's law**. Symbolically Hooke's law states that

$$T = \frac{\lambda x}{l},$$

where $\lambda$ is known as the modulus of elasticity of the string and $l$ is the unstretched (natural) length of the string. The ratio $\frac{\lambda}{l}$ is called the stiffness of a string, though it is more often used in the context of springs rather than strings.

**Springs** have all the properties of elastic strings but can be compressed as well as extended and in this case exert a force (thrust) away from their ends. This thrust still satisfies Hooke's law with $x$ now denoting the compression.

**Thin rods (or beams)** have all the properties of a string but can exert either a thrust or a tension at their ends. They are assumed to be rigid and, unlike a string, can sustain a force perpendicular to their lengths.

**Smooth surfaces** exert a reaction perpendicular to, and away from, themselves as shown in the two left hand diagrams below. The reaction of a smooth surface on a body cannot be negative. A zero reaction means that contact is about to be broken.

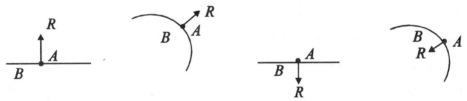

If there are two smooth bodies $A$ and $B$ in contact then the reaction of $A$ on $B$ is equal in magnitude and opposite in direction to that of $B$ on $A$ as shown in the two right hand diagrams above.

This is **Newton's third law.**

**Smooth pegs** The reaction at such pegs is normal to the peg and if a string is passed over such a peg, as shown in the diagram under Strings, then the tensions $T_1$ and $T_2$ are equal.

**Smooth pulleys** have, as far as Statical problems are concerned, the same properties as smooth pegs.

**Simply (smoothly) supported rods (beams)** The reactions of the supports on the rod or beam are perpendicular to it as shown below.

*Forces acting at a point*

**Rough surfaces** The reaction on a body in contact with such surfaces is not normally perpendicular to the surface.

As well as the perpendicular (or normal) component of reaction $R$ there is a force along the surface, the friction force of magnitude $F$ as shown in the left hand diagram below.

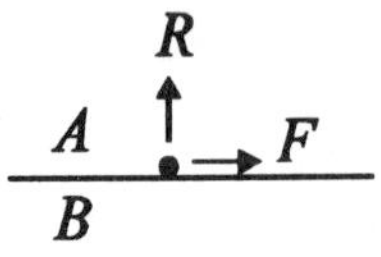

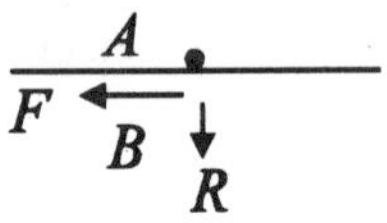

This force acts so as to oppose a tendency to move so that in the above diagram the motion of $A$ would be to the left and $F \leq \mu R$, where $\mu$ is the coefficient of friction.

If a body is on the point of sliding on another then the friction is said to be limiting and $F = \mu R$.

Newton's third law holds also for rough surfaces as shown in the right hand diagram above.

In solving problems the first step is to use the Glossary to interpret the problem and make the correct modelling assumptions to determine the type of force acting at each point. For strings, it is particularly important to mark the tensions at both ends of each straight part of each string. It should be remembered that, for a light string, the tension is constant throughout a given string but need not be the same in two strings tied to the same point.

All the forces acting on all bodies should be marked clearly on a **force diagram**. Making a clear force diagram is an essential first step in any problem solving. In all previous cases you have been presented with force diagrams but now you have to make your own.

If there are two bodies $A$ and $B$ in contact then there will be a force diagram for each and in making the force diagram you have to use Newton's third law which states that the force of $A$ on $B$ is equal and opposite to that of $B$ on $A$.

The problems can then be solved as before by resolving in two different directions. You should remember that a good choice of directions in which to resolve can simplify your calculations as you saw in Example 2.9.

The problems that you can solve are those with forces acting at a point. There may be several points connected in some way and you will have to look at each point separately.

It is easier to start with problems not involving friction.

*Forces acting at a point*

## Problems not involving friction

### Example 2.10
A particle of weight $W$ is suspended in equilibrium from the end $B$ of a light inextensible string $AB$. Find the tension in the string and the force necessary to hold the string at $A$.

Since the string is light the tension will be constant throughout it and, as shown in the force diagram, will be acting inwards from the ends. There will be the force due to gravity of magnitude $W$ acting vertically downwards at $B$. Nothing is known about the force at $A$ and therefore it is safest to assume that it has a vertical component $Y$ and a horizontal component $X$. All these forces are shown in the force diagram.

The forces acting at all points are in equilibrium. The only forces acting at $B$ are the tension acting directly upwards and the weight acting directly downwards and therefore
$$T = W.$$
Looking next at $A$, the only horizontal component of force is $X$ and therefore for equilibrium $\qquad X = 0.$

The component of force acting vertically upwards is $Y$ and that downwards is $T$ and therefore for equilibrium
$$Y = T = W.$$

### Example 2.11
A particle of mass $m$ is suspended in equilibrium from the lower end $B$ of a light elastic string $AB$, the upper end being held fixed. The elastic modulus of the string is $10mg$ and its unstretched length is $a$. Find, assuming that Hooke's law holds, the length of the string in the equilibrium position.

27

The forces acting are the same as in Example 2.10, and therefore the same diagram can be used though, since the mass is given rather than the weight, $W$ should be replaced by $mg$. The equilibrium of the particle gives
$$T = mg$$
Since Hooke's law holds
$$T = \frac{10mgx}{a},$$
where $x$ is the extension. Substituting $T = mg$ gives $x = \frac{a}{10}$ so that the total length of the string is $\frac{11a}{10}$.

**Example 2.12**

A small smooth ring $R$ of weight $W$ is threaded on a light inextensible string of length $8a$. The ends of the string are attached to two points $A$ and $B$ in a horizontal line and at a distance $2a$ apart. The system is in equilibrium in a vertical plane. Find the tension in the string.

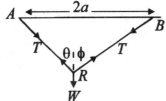

The string is light and therefore the tension will be constant on both the straight parts of the string. The ring is smooth so that the tension is the same on both parts of the ring and therefore the forces acting are as shown in the above force diagram.

There is no reason to assume immediately that both strings are inclined at the same angle to the horizontal, though symmetry suggests this, and therefore it is safer to assume that the angle between the string and the vertical is $\theta$ to the left of the string and $\phi$ to the right. Resolving horizontally the forces on the ring gives
$$T \sin \theta = T \sin \phi,$$
so that $\theta = \phi$ and therefore $AR = BR = 4a$ and $R$ is directly below the middle point of $AB$. Resolving vertically the components of the forces at the ring gives
$$2T \cos \theta = W.$$
The depth of $R$ below $AB$ is, by Pythagoras' Theorem, $\sqrt{15}a$ so that $\cos \theta = \frac{\sqrt{15}}{4}$.
Therefore,
$$T = \frac{2W}{\sqrt{15}}.$$

## Example 2.13

Two light inextensible strings are each tied to a particle of mass $m$.
The other ends are attached to two points in a horizontal line so that the particle is in equilibrium with the strings inclined at angles of 30° and 60° to the horizontal. Find the tensions in the strings.

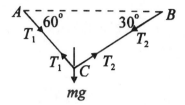

The force diagram is shown above. Since the strings are tied to the particle the tensions in them could be different and will be denoted by $T_1$ and $T_2$ respectively. In this case the mass of the particle is given rather than the weight so that the force due to gravity is $mg$.

There are two unknowns and they can be found by equating components in two different directions. The most obvious choice is to equate components vertically and horizontally, this would give two equations involving $T_1$ and $T_2$ and these have then to be solved for $T_1$ and $T_2$.

An alternative is to consider components along the two strings, these are perpendicular and therefore, for example, the tension along $AC$ does not have a component along $BC$. The problem is very similar to that of Example 2.9.

The component of the force of gravity along $AC$ is $mg \cos 30°$ so

$$T_1 = mg \cos 30° = \frac{\sqrt{3}mg}{2}.$$

The component of the force of gravity along $BC$ is $mg \sin 30°$ so that

$$T_2 = mg \sin 30° = \frac{mg}{2}.$$

## Example 2.14

Two small particles, each of weight $W$ are attached to the ends of a light inextensible string. The string passes over a small smooth peg and the particles are in equilibrium in a vertical plane. The string, at the points where it loses contact with the peg, is vertical. Find the force exerted on the peg.

Since the peg is smooth the tension on both sides of the peg is the same and acts vertically down and the force diagram is :-

*Forces acting at a point*

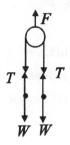

The strings therefore do not exert a horizontal force on the peg.
Resolving vertically for either particle gives
$$T = W.$$
The only force exerted on the peg by the string is $2T = 2W$ acting downwards.

### Example 2.15
The pulley system shown below is used to support a crate of mass 150 kg. Find, assuming the pulleys are smooth and light, and the rope is light and modelling the crate as a particle, the force that has to be applied at the end of the rope to maintain equilibrium.

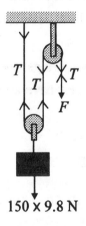

The forces acting are shown in the diagram. Since the pulleys are smooth the tensions in all parts of the rope are the same. The equilibium of the crate gives
$$2T = 150 \times 9.8 \text{ N},$$
so that $T = 735$ N. Equilibrium at the end of the rope gives
$$F = T,$$
so that the required force is 735 N.

## Exercises 2.4

In numerical examples, forces should be found to three significant figures and $g$ should be taken as 9.8 ms$^{-2}$.

**1** A small body is suspended in equilibrium from a fixed point by a light inextensible string. Find

(a) given that mass of the body is 0.4 kg, its weight and the tension in the string,

(b) given that weight of the body is 14.7 N, its mass and the tension in the string.

**2** A small parcel is placed on a horizontal table. Modelling the parcel as a particle find

(a) the reaction perpendicular to the table when the particle is of mass 3 kg,

(b) the mass of the parcel given that the normal reaction is 19.6 N.

Questions 3 to 5 refer to a body of weight $W$ N suspended from a fixed point by a light elastic string of natural length $a$ m, elastic modulus $\lambda$ N. The extension is denoted by $x$ m.

**3** Find $x$, given $W = 21$, $a = 2$, $\lambda = 105$.

**4** Find $\lambda$, given $W = 50$, $x = 0.2$, $a = 4$.

**5** Find $a$, given $W = 30$, $x = 0.2$, $\lambda = 210$.

Questions 6 and 7 refer to a particle of mass 0.4 kg suspended by a light inextensible string, the other end of which is attached to a fixed point.

**6** The particle is acted on by a horizontal force so that it is in equilibrium with the string inclined at an angle of 40° to the downward vertical. Find the force.

**7** The particle is maintained in equilibrium with the string inclined at an angle of 30° to the downward vertical by a force acting on the particle perpendicular to the string. Find the force and the tension in the string.

**8** A particle of mass 0.3 kg is suspended by two light inextensible strings from two fixed points on the same horizontal level. The strings are inclined at angles 25° and 35°, respectively, to the horizontal. Find the tensions in the strings.

**9** A smooth ring $R$ of mass $m$ slides on a light inextensible string whose ends $A$ and $B$ are fixed at two points on the same level. A horizontal force of magnitude $P$ is applied at $R$ so that the ring is in equilibrium vertically below $A$ with $BR$ inclined at an angle $\alpha$ to the vertical. Find $P$.

**10** Two identical light elastic strings $AB$ and $BC$, each of natural length 0.8 m and modulus 3000 N, are joined together at $B$. A particle of weight 900 N is attached to $C$ and is suspended in equilibrium by the composite string $ABC$ with the end $A$ fixed.
Find the length $ABC$.

11 Assume now that in question 10 the strings are not joined to each other at $B$ but both attached to a particle of weight 20 N with the particle being between the strings. Find the length $ABC$ when the particle of weight 900 N is now suspended in equilibrium.

12

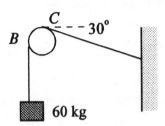

The diagram shows a crate of mass 60 kg supported by a rope passing over a small pulley. The other end of the rope is attached to a fixed point. The pulley is circular and the rope just loses contact with the peg at $B$ and $C$. The rope is vertical at $B$ and at an angle of 30° to the horizontal at $C$. Neglecting the weight of the rope, modelling the crate as a particle and assuming the pulley to be smooth, find the tension in the rope and the horizontal and vertical components of the force acting on the pulley.

13

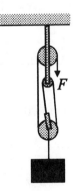

Find the force, $F$, that has to be applied to the end of the rope in order that the pulley system can support the box of weight 80 kg. The pulleys are smooth and their mass and that of the ropes may be neglected. Also all parts of the string may be assumed to be vertical.

14 A light elastic string $AB$, of natural length 1.5 m and modulus 200 N, has the end $A$ fixed and a heavy particle attached to the end $B$. A horizontal force of magnitude $F$ is then applied at $B$ so that the system is in equilibrium with $AB$ taut and inclined at an angle of 30° to the downward vertical with $AB = 1.8$ m. Find the value of $F$ and the mass of the particle.

**15** Two light strings, attached to a particle $P$ of mass $M$, pass over two smooth pegs at the same level and hang vertically in equilibrium with masses $3m$ and $4m$ at their ends. Given that the strings at $P$ are perpendicular to each other, find the ratio $\dfrac{m}{M}$.

**16**

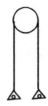

The diagram shows a light string, with a scale pan attached at each end, passing over a small peg. The mass of each scale pan is 0.1 kg and the system is initially in equilibrium with 0.9 kg in each scale pan and the string at the points where it just loses contact with the peg is vertical. Find the tension in the string.

It was found that a further 0.4 kg could be placed in one scale pan before equilibrium was broken. Find the tensions in the two parts of the string when equilibrium is about to be broken.

It may be assumed that motion did not take place immediately when the mass in one scale pan was increased because the peg is rough. A model taking into account the roughness of the peg shows that when the string is about to move the tension increases along the string in the direction in which motion would occur so that the ratio of the tensions at the two points at which the string is about to lose contact with the peg is $e^{\mu\pi}$, where $\mu$ is the coefficient of friction. Use this model to estimate the coefficient of friction.

## Problems involving friction

The simplest types of problems involving friction are those when the frictional force necessary for equilibrium has to be found and these are essentially the same as the problems you have already solved.

The next class, in order of difficulty, is that when equilibrium is on the point of being broken and in these cases the equilibrium is limiting so that, provided the direction of friction is known, a force diagram can be drawn which includes the frictional forces. The problem can then be solved by resolving in two different directions.

In all other cases the best tactic is to form a force diagram showing arbitrary values of the friction force $F$ and the normal reaction $R$. Expressions for $F$ and $R$ can be obtained by resolving in the usual way and then the condition $F \le \mu R$ applied. The only snag is that you may not always have chosen the correct direction for friction and "your" $F$ may not be the magnitude. If this happens then you will be probably end up with something which is automatically true or nonsense and no progress will have been made. You can avoid this by using $-\mu R \le F \le \mu R$. It is particularly important that in questions where an inequality is required that you use $F \le \mu R$ (or $-\mu R \le F \le \mu R$) rather than assume friction is limiting and then put in an inequality on the last line. If the inequality is not given then there is a 50% chance of you getting the inequality the "wrong way"!

### Example 2.16

A particle of weight 5 N is in equilibrium on a rough plane inclined at an angle of 60° to the horizontal. Find the normal reaction and the friction force acting on the particle.

The force diagram is :

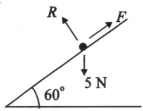

For problems involving inclined planes it is often easier to take components along and perpendicular to the plane rather than horizontally and vertically. The components of the weight are 5 sin 60° N down the plane and 5 cos 60° N perpendicular to the plane and in the sense away from the particle into the plane.

Therefore resolving along and perpendicular to the plane gives

$R = 5 \cos 60°$ N $= 2.5$ N   and $F = 5 \sin 60°$ N $= 2.5 \sqrt{3}$ N.

# Forces acting at a point

The direction of $F$ was taken to be up the plane since any motion would be downwards. If the wrong direction had been chosen then this would have shown up in a negative value of $F$.

## Example 2.17

A particle of weight 8 N is at rest on a rough horizontal table, the coefficient of friction between the particle and the table is 0.4. When a horizontal force of magnitude $P$ is applied to the particle it is just about to slide. Find the value of $P$.

The force diagram is:

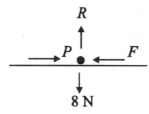

Resolving vertically gives $\qquad R = 8$ N.

In this case the particle is about to move so that friction is limiting and therefore
$$F = \mu R = 3.2 \text{ N}.$$
Resolving horizontally gives $\qquad P = F = 3.2$ N.

## Example 2.18

A particle $P$ of mass $6m$ lies on a rough horizontal table, with coefficient of friction $\frac{1}{4}$.

The particle is attached by a light inextensible string to a second particle $Q$. The string passes over a smooth pulley at the edge of the table and $Q$ hangs in equilibrium. Find the greatest possible mass of $Q$.

The force diagram is :

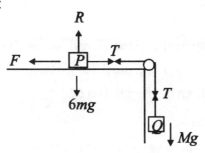

It is assumed that $Q$ has mass $M$. If motion were to occur it would be to the right and therefore the friction force is to the left. The only forces acting on $Q$ are vertical and, resolving vertically for $Q$ gives
$$T = Mg.$$

Resolving horizontally for $P$ gives
$$F = T,$$
so that
$$F = Mg.$$
Resolving vertically for P gives
$$R = 6mg.$$
Therefore
$$\frac{F}{R} = \frac{M}{6m}$$
The maximum value of this ratio is $\frac{1}{4}$ so that $\frac{M}{6m} \leq \frac{1}{4}$.

So the greatest possible mass of $P$ is $\frac{3m}{2}$.

## Example 2.19

A heavy particle of weight $W$ is placed on a rough plane inclined at an angle $\alpha$ to the horizontal. The coefficient of friction between the plane and the particle is $\mu$. Show that equilibrium is not possible unless $\tan \alpha \leq \mu$.

Find, when $\tan \alpha > \mu$, the least value of the magnitude of a force acting up the line of greatest slope of the plane which will maintain equilibrium.

The force diagram for the first part is:

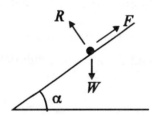

Since the particle is likely to slip down the plane $F$ will act up the plane and resolving along and perpendicular to the plane as in Example 2.16 gives
$$F = W \sin \alpha, \quad R = W \cos \alpha,$$
and
$$\frac{F}{R} = \tan \alpha,$$
Therefore $\tan \alpha \leq \mu$ for equilibrium.

In the second part the force of magnitude $P$ acts up the plane. The solution to the first part shows that for $\tan \alpha > \mu$ the particle would slip down the plane, therefore the least force to obtain equilibrium is that which just stops the particle sliding down. The friction force will still act up the plane.

The force diagram is therefore:

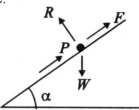

Resolving along and parallel to the plane gives
$$F + P = W \sin \alpha, \quad R = W \cos \alpha.$$
The condition $F \le \mu R$ gives
$$W \sin \alpha - P \le \mu W \cos \alpha$$
and therefore
$$W \sin \alpha - \mu W \cos \alpha \le P.$$
Therefore the least value of $P$ is $W(\sin \alpha - \mu \cos \alpha)$.

It is a good idea in examples like this involving friction to check that the answer is sensible. Here it is in the sense that, since $\tan \alpha > \mu$, it is at least positive!

### Example 2.20
Find for the problem in Example 2.19, when $\tan \alpha > \mu$, the greatest value of the magnitude of a force acting up the line of greatest slope of the plane which will maintain equilibrium.

The greatest value of $P$ would be when the particle is about to slip up the plane and $F$ is acting downwards. If $P$ is assumed to still act in the same direction then the same equations as before are found. Applying the condition $-\mu R \le F \le \mu R$ gives
$$\mu W \cos \alpha \le W \sin \alpha - P \le \mu W \cos \alpha,$$
so that
$$W \sin \alpha - \mu W \cos \alpha \le P \le W \sin \alpha + \mu W \cos \alpha.$$
The left hand inequality gives the least value of $P$ and the right hand one the maximum value. The maximum value occurs when the particle is about to move up the plane so friction is acting downwards and the force has to overcome both friction and the force of gravity.

### Example 2.21
Two small rough rings $A$ and $B$, of weights $3W$ and $W$ respectively, slide on a fixed, rough, horizontal rod. The coefficient of friction between each rod and the ring is 0.5. A light inextensible string is threaded through a smooth ring of weight $W$ and its ends are attached to $A$ and $B$. The whole rests in equilibrium in a vertical plane. Explain why both parts of the string are inclined at the same angle $\theta$ to the vertical and find the greatest value of $\theta$ for which equilibrium is possible.

The force diagram is

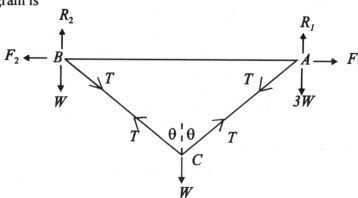

If the strings were inclined at different angles $\theta$ and $\phi$ then resolving horizontally at $C$ would give
$$T \sin \theta = T \sin \phi,$$
so that
$$\theta = \phi.$$

This is a fairly complicated problem in that it is necessary to resolve at $A$, $B$ and $C$.
Resolving vertically at $C$ gives
$$2T \cos \theta = W.$$
Resolving horizontally and vertically at $A$ gives
$$T \sin \theta = F_1, \quad R_1 = 3W + T \cos \theta.$$
Resolving horizontally and vertically at $B$ gives
$$T \sin \theta = F_2, \quad R_2 = W + T \cos \theta.$$
Substituting for $T \cos \theta$ gives
$$R_1 = \frac{7W}{2}, \quad R_2 = \frac{3W}{2} \text{ and } F_1 = F_2 = \frac{W \tan \theta}{2}.$$
The friction forces are the same at both $A$ and $B$ and since $R_2 < R_1$ it follows that slipping will first occur at $B$ when
$$\frac{F_2}{R_2} = \frac{1}{3} \tan \theta \leq 0.5.$$
The maximum value of $\tan \theta$ is 1.5 giving the maximum value of $\theta$ as approximately 56.3°.

### Exercises 2.5

In numerical examples, $g$ should be taken as 9.8 ms$^{-2}$ and answers given correct to three significant figures.

1 A particle is in equilibrium on a rough horizontal table. A string is attached to the particle and is inclined at an angle of 40° to the horizontal. $T$ denotes the tension in the string and $F$ the friction force. Find
(a) $F$ given that $T = 40$ N, (b) $T$ given that $F = 60$ N.

**2** A particle of mass 1.5 kg, at rest on a rough horizontal plane, can just be moved by a horizontal force of magnitude 5 N. Find the coefficient of friction.

**3** A particle of mass 2 kg is in equilibrium on a rough horizontal plane, the coefficient of friction being 0.4. Find the least force which, acting (i) horizontally, (ii) at an angle of 30° to the upward vertical, would just move the body along the plane.

**4** A particle of mass 3 kg is placed on a rough plane inclined at an angle of 50° to the horizontal. The coefficient of friction between the plane and the particle is 0.25. Find the least force acting along a line of greatest slope of the plane required (i) to prevent the particle from sliding down, (ii) to move it up the plane.

**5** A particle of weight 80 N is held in limiting equilibrium on a plane inclined at an angle of 30° to the horizontal by a horizontal force. Given that the coefficient of friction is 0.4 find the magnitude of the force when

(i) the particle is about to slip up the plane,

(ii) the particle is about to slip down the plane.

**6** A particle of mass 3kg is on the point of sliding down a rough plane inclined at an angle $\alpha$ to the horizontal when a force of magnitude 5 N is applied up the plane along a line of greatest slope. When the force is increased to 10 N the particle is on the point of moving up the plane. Find $\sin \alpha$.

## 2.3 Calculation of the resultant of forces acting at a point

The main purpose of calculating the resultant is for use in problems involving motion but the idea of a resultant is also handy in equilibrium problems. For example, if there are two forces acting and a third has to be included to produce equilibrium then the third force would have the same magnitude as the resultant of the other two but act in the opposite direction. This could occur with wires at the top of a telegraph pole when two had been attached and the third had to be placed so as to give equilibrium.

The components of the resultant of several forces acting at a point are the sum of the components of the individual forces. Therefore if the resultant of a force can be calculated from its components then the resultant of any number of forces can be found.

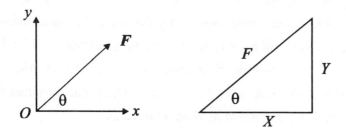

## Forces acting at a point

The components of a force $F$ inclined at an angle $\theta$ to the $x$ direction as in the diagram are
$$X = F \cos \theta, \quad Y = F \sin \theta.$$

If $\theta$ is acute then $F$ is the hypotenuse of a right angled triangle whose other sides are $X$ and $Y$ and therefore by Pythagoras' theorem
$$F = \sqrt{X^2 + Y^2}.$$
It is actually possible to prove that this is true even if $\theta$ is not acute (so that one or both of $X$ and $Y$ may be negative) and from now on it will be assumed to be true for all $X$ and $Y$. Therefore the magnitude of the resultant can be calculated easily.

Finding the direction is a bit trickier. Dividing the components gives
$$\tan \theta = \frac{Y}{X}.$$

You have met this kind of equation before in working with right angled triangles. When $X$ are $Y$ both positive, $\cos \theta$ and $\sin \theta$ are also positive so that $\theta$ is acute and you can find it by using the $\tan^{-1}$ function on your calculator. When one of $X$ or $Y$ is negative it is necessary to be more careful and the first step is to find out in which quadrant the line representing the resultant lies. The four possibilities are shown below.

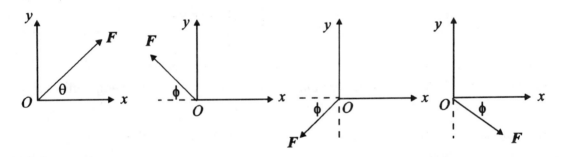

The acute angle $\phi$ between the resultant and the $x$ axis can be found from
$$\tan \phi = \frac{|Y|}{|X|},$$
i.e. you drop the minuses on the components. This lets you work out the exact position of the resultant and you can then work out the angle with the positive $x$-direction or any other line.

It is possible to carry out the calculation more directly by finding $F$ first and using the $\cos^{-1}$ or $\sin^{-1}$ functions to find $\theta$. You still have to determine the quadrant in which $\theta$ lies and use the symmetries of the trigonometric functions (for $\theta$ in degrees these are $\sin(180 - \theta) = \sin \theta$, $\cos(-\theta) = \cos \theta$, $\tan(180 + \theta) = \tan \theta$). Until you have had more practice in trigonometry you may find this second method a bit tricky.

*Forces acting at a point*

In working out resultants you can choose any two perpendicular reference directions that you like but for two forces you could simplify things by taking one reference direction parallel to one of the forces.

**Example 2.22**

Find the resultant when (a) $X = 5$ N, $Y = 2$ N, (b) $X = -4$ N, $Y = 4$ N (c) $X = -6$ N, $Y = -3$ N, (d) $X = 3$ N, $Y = -1$ N.

(a) This is the simplest case, corresponding to diagram (a) above, so that

$$R = \sqrt{5^2 + 2^2} \text{ N} = 5.39 \text{ N, also } \tan\theta = \frac{2}{5} \text{ so that } \theta = 21.8°.$$

(b) This corresponds to diagram (b) above where the line representing the resultant lies in the second quadrant.

The resultant is therefore $\sqrt{4^2 + 4^2}$ N = 5.67 N, $\tan\phi = \frac{4}{4}$ so $\phi = 45°$ and therefore the resultant is at an angle of 135° to the positive *x*-direction.

(c) This corresponds to diagram (c) above with the line representing the resultant lying in the third quadrant.

The resultant is therefore $\sqrt{6^2 + 3^2}$ N = 6.71 N, $\tan\phi = \frac{1}{2}$ so $\phi = 26.6°$ and therefore the resultant is at an angle of 206.6° to the positive *x*-direction.

(d) This corresponds to diagram (d) with the line representing the resultant being in the fourth quadrant.

The resultant is therefore $\sqrt{3^2 + 1^2}$ N = 3.16 N, $\tan\phi = \frac{1}{3}$ so $\phi = 18.4°$ and therefore the resultant at an angle of $-18.4°$ (or 341.6°) to the *x* direction.

When finding the resultant for several forces acting there will be an additional step of adding the separate components to get the components of the resultant force.

**Example 2.23**

The diagram shows three forces *P*, *Q* and *R* acting at a point. The magnitudes (in newtons) are shown in brackets. Find the magnitude and direction of their resultant.

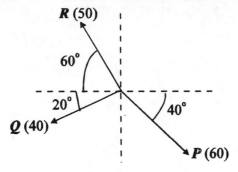

The components of *P*, *Q* and *R*, respectively, to the right across the page are (in newtons) 60 cos 40°, − 40 cos 20° and − 50 cos 60°. The component of the resultant is the sum of these which is −16.6 N.

The components of *P*, *Q* and *R*, respectively, up the page are (in newtons) − 60 sin 40°, − 40 sin 20° and 50 sin 60°. The component of the resultant is the sum of these which is − 8.95 N.

The resultant is of magnitude $\sqrt{16.6^2 + 8.95^2}$ N = 18.9 N, the line representing the resultant lies in the third quadrant. The tangent of the acute angle between the direction of the resultant and that of the dotted line is $\frac{8.95}{16.6}$ = 0.539, so this angle is 28.3°. Therefore the resultant makes an angle of 208.3° with the line to the right and across the page.

## Example 2.24

The three forces in the previous example model the forces in three horizontal wires at the top of a telegraph pole. Find the position of, and tension in, a fourth horizontal wire to be placed so that the four forces will be in equilibrium.

The fourth wire will be in the opposite direction to the resultant of the other three, i.e. it acts at an angle of 28.3° to the line to the right and across the page and the tension in this wire is 18.9 N.

## Exercises 2.6

In numerical examples answers should be given correct to three significant figures.

**1** The *x*- and *y*- components of a force are denoted by *X* and *Y*. Find the magnitude and direction, referred to the positive *x*-direction, of the resultant when

(a) *X* = 7 N, *Y* = 3 N,    (b) *X* = 4 N, *Y* = −8 N,    (c) *X* = −3N, *Y* = 11 N,    (d) *X* = −5 N, *Y* = −1 3 N,    (e) *X* = −7 N, *Y* = 4 N,    (f) *X* = −3 N, *Y* = −3 N.

**2** The diagram shows three forces *P*, *Q* and *R* acting at a point *O*. Find the magnitude of their resultant, and the direction it makes with the positive *x*- direction, when

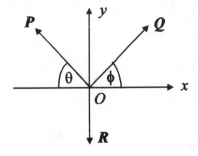

(a) $P = 1$ N, $Q = 2$ N, $R = 5$ N, $\theta = 20°$, $\phi = 30°$,
(b) $P = 8$ N, $Q = 3$ N, $R = 4$ N, $\theta = 40°$, $\phi = 60°$,
(c) $P = 6$ N, $Q = 6$ N, $R = 1$ N, $\theta = 50°$, $\phi = 20°$.

**3** Find the additional force that will have to be introduced into each of the cases in the previous exercise in order that the system of four forces is in equilibrium.

## 2.4 Modelling assumptions

### Force of gravity

You know that if you release anything just above the surface of the earth then it will drop down. There is therefore a force acting on it. This is the force of gravity acting on the body. It acts towards the centre of the earth and its magnitude is the weight of the body.

The weight of the body can also be expressed in terms of the mass, $m$, of the body and $g$, the acceleration due to gravity. These quantities will be defined for you more precisely later but for the time being all you need to know is that, for any body, there is a precisely defined quantity called its mass and that $g$ is approximately $9.81$ ms$^{-2}$. The unit of mass is the kilogram (abbreviation kg). The weight in newtons of a body of mass $m$ kg is $mg$ where $g$ is measured in ms$^{-2}$.

The actual force exerted by gravity varies in magnitude with the distance from the earth's centre and also varies with latitude. In most circumstances these variations are ignored and the usual modelling assumption is that the weight of a particle is constant and that the force of gravity acts along the vertical.

For the simple model of a body as a particle the force of gravity acts at the point occupied by the particle but for a general body, which is a collection of particles, the situation is not so clear. There is however, for any body, a unique point through which the force of gravity acts, this point is known as the centre of gravity of the body. You will not be expected to know the position of the centre of gravity of any particular body except that the centre of gravity of a thin uniform rod lies at its midpoint.

One device for measuring weight is the spring balance, another is the ordinary bathroom scales. The actual reading on the scales gives the magnitude of the reaction of the scales on the body and, when the scales are not moving, this is equal to the weight of the body.

These devices do not actually measure the true weight (i.e. the magnitude of the force of gravity) since there should be a slight correction due to the effect of the rotation of the earth. This is normally neglected but that in itself is a modelling assumption.

A body without weight (more precisely one where the gravitational force acting on it may be neglected compared to other forces) is called light.

You should be careful to notice that there is a difference between being without weight and being weightless, which is something you may have heard of. Weightlessness is normally associated with motion and a body is said to be weightless if there is no reaction between it and a surface it is in contact with. For example if someone were unlucky enough to be standing on bathroom scales in a lift which was falling freely under gravity then the scale would not show a reading. Similarly if you were falling freely and holding a suitcase then you would not feel the weight of the suitcase and so to you it would appear weightless (see Example 5.3). Even these statements are not precisely correct because they assume that there are no other forces than gravity acting, i.e. there are no resistances. This would mean that these statements would only be valid in a vacuum.

**Strings**

A small body $P$ is attached to one end $A$ of a string $AB$, the other end $B$ is held fixed so that the body hangs at rest as shown in diagram (a) below.

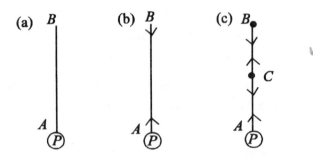

To simplify matters it will be assumed that $P$ is modelled as a particle. There is the force of gravity acting vertically downwards on $P$ and therefore, since it is at rest (i.e. in equilibrium) there is an equal and opposite force acting in the string from $A$ to $B$. Such a force directed from one end to the other is called the tension. The individual holding the string at $B$ would experience a force acting in the direction of $B$ to $A$. Therefore, a string exerts a pull (tension) at its extremities as shown in diagram (b) above. For any point $C$ between $A$ and $B$, then the part $AC$ will exert some force on the part $BC$ and this force will be in the direction $CA$ and the part $BC$ will exert a force on $AC$ in the direction $CB$. The situation is as shown in diagram (c).

Therefore, a string sustains a tension at all points along its length and the tension is in fact the force of interaction between two parts of the string.

The word string in Mechanics implies something which has length but no cross section so that it can be modelled by a thin line or curve. A string can only pull from its ends and not push. It is also assumed that it is perfectly flexible i.e. it cannot exert a force perpendicular to its length.

## Forces acting at a point

If the string in the above diagrams is assumed light then there will be no force due to gravity acting at any point of the string and therefore the force exerted by the part *BC* will be equal in magnitude to that exerted by the part *AC*. Therefore the tension will be constant. It is possible to prove that assuming a string is light means that the tension will be constant throughout its length and that, if any two points of a light string are held fixed, the string in the region between them will be straight.

Cables and ropes of all kinds are modelled as light strings and this need not always be an adequate model. For example the left hand diagram below shows a tug just attached to a ship and the right hand one shows the tug about to move the ship.

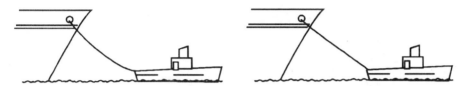

A light string is obviously not a good model in the left hand case since the rope is sagging but it might be reasonable in the second case. The difference is that in the second case there is considerable tension in the rope and that tension is considerably greater than the weight of the rope. Therefore modelling a rope by a light string is effectively ignoring the weight of the rope relative to the tension acting on it.

### Elastic strings

You know that if you pull at the two ends of a piece of string of the type used to tie up parcels then, as far as you can see, the distance between the two ends will stay the same. If, on the other hand, you pulled the two ends of a thin piece of elastic then the distance between the ends will increase. One way of distinguishing between these is to say that the parcel string is inextensible and that the piece of elastic is extensible. The difference is however more than this and modelling elastic strings is not always straightforward.

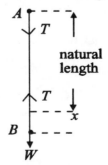

The diagram shows an elastic string *AB* suspended from a fixed point *A* and a weight *W* is attached to the end *B*. The length of *AB* can be found and the extension of the string determined. (Determination of the extension requires knowing the unstretched, or natural, length of the string. This can be found by putting the string so that it is just straight on a table and measuring its length). By putting different weights at $B$ the extensions corresponding to different weights can be found. Since the weights are in equilibrium the tension in the string is equal to the weight. Therefore the tensions corresponding to different extensions can be found. Experiments of this kind have been carried out for many materials and the graph of the tension $T$ against extension $x$ is roughly as shown below.

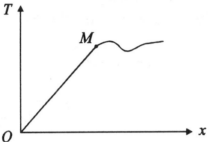

Up to a certain value of $T$ the graph will be a straight line $OM$, thereafter it may take on a more complicated form, one possibility is shown in the above diagram. For sufficiently large tensions the string eventually breaks.

For tensions corresponding to the part $OM$ of the graph, if the weights are removed then the string regains its natural length and if the experiment is repeated then the line $OM$ will again be obtained. If the weights are removed for tensions greater than those corresponding to the part $OM$ then the string will not regain its natural length and the same graph will not be obtained on repeating the experiment. The tension corresponding to the point $M$ represents the elastic limit of the string.

A light extensible (elastic) string possesses some of the properties of a light inextensible string in that it can only sustain a tension (that is, it cannot push) and the tension is constant along its length. It differs in that, as the adjective extensible suggests, the length of the string is not constant.

The usual model of an elastic (extensible string) is that it possesses all the attributes of a string but that the tension is described by the line $OM$ i.e. the tension is directly proportional to the extension. This is Hooke's law which states that when an elastic string of natural length $l$ is extended by an amount $x$ then the magnitude of the tension $T$ in the string is given by
$$T = \frac{\lambda x}{l}$$
where $\lambda$ is an experimentally determinable constant known as the modulus of elasticity of the particular string, while $\lambda/l$ is often called the stiffness and also the string constant.

**Springs** A spring is effectively a spiral of thin wire and it is usually assumed that it can be modelled as an elastic string though it has one additional property in that it can sustain a thrust as well as a tension. This means that a spring can be compressed and when compressed there is a thrust acting outwards at the ends.

Hooke's law still applies to a spring though, for a compression, the tension becomes a thrust.

**Rods** A rod, like a string, is modelled as a line. Rods are assumed to be rigid and inflexible so that they are modelled by straight lines. There are two assumptions made, using a rod model, which are not valid for a string model. One is that a rod can exert thrusts as well as tensions, the other is that a rod can sustain forces perpendicular to itself. This means that a rod can be supported by forces applied at two points of itself as shown in the diagram and can be used to give, for example, a simple model of a plank across a river or even of a bridge. In reality most planks, if their weight cannot be neglected, will bend slightly if supported as above and are slightly elastic. The rod model ignores this possibility.

**Smooth surfaces**

If a small body $A$ is in contact with a surface $S$ as shown below in diagram (a)

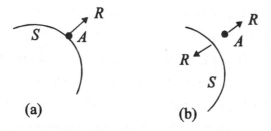

there will be some force exerted by $S$ on $A$. A smooth surface is defined to be such that its reaction on $A$ is normal to $S$ at the point of contact and in the direction from $S$ to $A$. There will also be, by Newton's third law, an equal and opposite force acting on $A$ due to $S$ and in some problems it is necessary to show the bodies as slightly separated as shown in diagram (b) above and the interactive forces on each body shown clearly. The reaction of the surface is always away from it and, if it becomes zero (or is inwards) in your calculations then this means that contact is about to be (or has been) lost.

If something like a book were pushed along a perfectly smooth table then only the slightest push would be necessary to move it and it would then continue to move without further effort. This is because a smooth surface would not exert a force tangential to itself. In reality no surface is perfectly smooth. If one did exist then it would not be possible to walk or drive on it. This is because the act of walking or driving exerts a tangential force on the road and, by Newton's third law, the road exerts an equal and opposite force on the foot or wheel and this is what makes motion possible.

**Smooth peg (or pulley)**  A peg or a pulley is used as shown below to change the

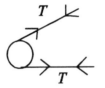

direction of a rope (modelled usually as a string). A smooth peg is one such that the only force that it exerts is perpendicular to its surface. Also if a string is passed round a smooth peg as shown above then the tension on both sides of the string is the same. In Statics a smooth pulley is modelled as a smooth peg. The modelling of a pulley, when motion is involved, is more complicated and is discussed in 5.3.

**Rough surfaces**

As mentioned above, surfaces, in reality, exert a tangential force on anything in contact with them. If you push a book on a table then initially you will experience some resistance, then the book suddenly slips and, if you are sufficently perceptive, you might notice that less effort is required to keep the book moving than was necessary to start it. Any surface which exerts a tangential (i.e sideways) force is said to be rough. The force at the point of contact with a rough surface has two components, the reaction $R$ perpendicular to the surface and a component $F$, the friction force. This is illustrated in the following diagram where the forces on both bodies in contact are shown, using Newton's third law.

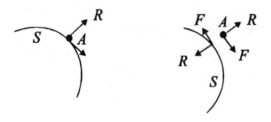

The following diagram illustrates an experiment that can be used to determine the behaviour at a point of contact with a rough surface.

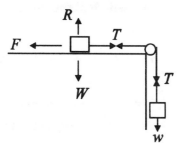

A small block of weight $W$ is placed on a table and a string attached to the block passes over a smooth pulley at the edge of the table and a weight $w$ is attached to the free end of the string so that the system is in equilibrium. The block is modelled as a particle, the only forces acting on it are its weight $W$ vertically downwards, the normal reaction $R$ acting vertically upwards, the tension $T$ in the string and the force of friction $F$ which is shown acting to the left. Resolving horizontally and vertically for the block gives

$$R = W, F = T,$$

and resolving vertically for the hanging weight gives

$$T = w.$$

The last equation assumes the pulley is smooth and that the string is light, it also shows since tension is positive that the correct direction was chosen for the force of friction.. Initially the values of $w$ can be increased, for given $W$, without disturbing equilibrium. At a particular value of $w$ the block just slips, and the values of $R$ ($= W$) and $F$ ($= w$) at slipping recorded. The process can be repeated for various values of $W$ and a graph of $R$ against $F$ (at slipping) drawn. The graph will be found to be of the form below, being initially a straight line and then curving.

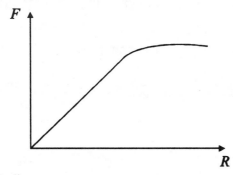

On the straight line part

$$F = \mu R,$$

and $\mu$ is called the coefficient of friction, or, more correctly, the coefficient of static friction and is a property of the two surfaces in contact. The normal model ignores the curved part of the curve and it will always be assumed that the relation between the normal reaction and the friction force at slipping is linear.

From experiments the following summarise the properties of the friction force at a rough surface.

(a) The force of friction acts in the sense so as to prevent motion of a body.

(b) Until the magnitude of the force of friction reaches a limiting value its magnitude is just sufficient to prevent slipping.

(c) At the limiting value $F = \mu R$ where $\mu$ is called the coefficient of friction.

(d) Until slipping occurs $F \leq \mu R$.

In many problems $F$ is used to denote the component in a particular direction and this may not always be in the direction in which the friction force is acting, therefore $F$ may be negative and therefore the correct condition for the component of the force of friction is
$$-\mu R \leq F \leq \mu R.$$
For $F = \mu R$ the friction is said to be limiting and a body is said to be in limiting equilibrium and is on the point of slipping.

Once a body has started slipping experiment shows that the friction force is still directly proportional to the normal reaction but the coefficient of proportionality is not always equal to $\mu$ and is sometimes denoted by $\mu'$ and is referred to as the coefficient of sliding friction. It is often found that $\mu' < \mu$, this explains the phenomenon of the book being harder to start sliding than to keep sliding and of a drawer which suddenly comes out with a rush.

In many cases the modelling assumption is that both the coefficients of friction are the same and, if you are not told specifically which coefficient of friction is given, then you should assume that you can use the same value for both limiting equilibrium and sliding. For steel on steel $\mu = 0.6$, $\mu' = 0.4$, for tyres on a dry road $\mu = 0.9$, $\mu' = 0.8$.

## Miscellaneous Exercises 2

**1** The components parallel to the *x*- direction of three forces acting at a point are 5 N, 7 N and 8 N. The components parallel to the *y*-direction of the forces are 4 N, 11 N and 6 N. Find

(a) the components in the *x*- and *y*- directions of the resultant of these forces,

(b) the magnitude of the resultant,

(c) the angle that the resultant makes with the *x*-direction,

(d) the magnitude and direction of the single additional force which will be in equilibrium with the other three forces.

**2** The components parallel to the *x*-direction of three forces acting at a point are 1 N, $-5$ N and $p$ N. The components parallel to the *y*-direction of two of the forces are 1 N, 3 N and the third has no component in the *y*-direction.

Given that the resultant of the forces has magnitude 5 N find the two possible values of $p$.

**3**

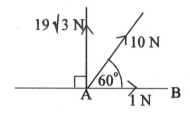

Three forces of magnitudes $19\sqrt{3}$ N, 10 N and 1 N act at the point $A$ in the directions shown in the above diagram. Find the magnitude of the single additional force acting at $A$ which will produce equilibrium and find the angle between this force and the line $AB$.

**4**

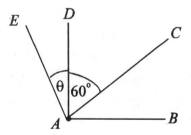

A particle of weight 18 N is hanging in equilibrium at a point $A$ supported by four strings $AB$, $AC$, $AD$, $AE$, all in the same vertical plane. The string $AB$ is horizontal and the tension in $AB$ is $7\sqrt{3}$ N; the string $AC$ makes an angle of $60°$ with the vertical and the tension in $AC$ is 8 N; the string $AD$ is vertical and the tension in $AD$ is 3 N; the string $AE$ makes an angle of $\theta°$ with the vertical and the tension in $AE$ is $T$. Find the values of $T$ and $\theta$.

**5** A particle $P$ of mass 0.2 kg is suspended in equilibrium from a fixed point $O$ by a <u>light extensible string of natural length</u> 0.4 m. State which one of the words underlined enables you to assume that the tension is the same at all points of the string.

Given that the modulus of elasticity of the string is 10 N find the distance $OP$.

**6** A particle of weight 60 N is attached to two inextensible strings each of length 13 cm. The other ends of the strings are attached to two points $A$ and $B$ on the same horizontal level at a distance of 24 cm apart. Find the tension in the strings when the particle hangs in equilibrium.

The inextensible strings are then replaced by elastic ones, each of natural length 13 cm and of the same modulus of elasticity. The particle then hangs in equilibrium 9 cm below the line $AB$. Find the modulus of elasticity of the strings.

**7** Forces of magnitude $P$ and $Q$ act along lines $OA$ and $OB$ respectively, and their resultant is a force of magnitude $P$; if the magnitude of the force along $OA$ is changed to $2P$ the resultant is again a force of magnitude $P$. Find

(i) $Q$ in terms of $P$,

(ii) the angle between $OA$ and $OB$,

(iii) the angles which the two resultants make with $OA$.

**8**

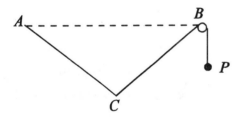

The diagram shows a light inextensible string fixed at a point $A$ and passing over a small smooth peg $B$ fixed at the same level as $A$. A particle $P$, of mass $m$, hangs freely from the other end of the string. A smooth ring $C$, also of mass $m$, is free to slide on the string between $A$ and $B$ and the system is in equilibrium. Show that $AC$ and $BC$ are both inclined at an angle $\pi/3$ to the vertical.

**9** A book of mass 1.5 kg rests on a rough plane inclined at an angle $\alpha$ to the horizontal. Given that the coefficient of friction between the plane and the book is 0.3 and that the book is on the point of slipping down the plane find $\alpha$.

**10** A car is parked on a hill with its brakes locked. The car is of mass 1200 kg and the hill is inclined at an angle of 20° to the horizontal. Find the total friction force and normal reaction of the road on the car.

What modelling assumption do you make in your calculations.

**11** The above car is then parked on a hill whose surface is such that the coefficient of friction between the tyres and the road is 0.6. Find the maximum slope such that the car will not slip down it.

**12** A block of mass 3 kg rests on a rough horizontal table. When a force of magnitude 10 N acts on the block at an angle of 60° to the horizontal in an upward direction the block is on the point of slipping. Find the coefficient of friction between the block and the table.

**13** A book is placed on a desk lid which is slowly tilted. Given that the book starts to slide when the lid is inclined at an angle of 30° to the horizontal, find the coefficient of friction.

**14** A particle is placed on a smooth plane inclined at an angle of 35° to the horizontal. The particle is kept in equilibrium by a horizontal force of magnitude 8 N acting in the vertical plane containing the line of greatest slope of the plane which passes through the particle. Find

(a) the weight of the particle,

(b) the magnitude of the force exerted by the plane on the particle.

**15** A particle is suspended in equilibrium by two light inextensible strings and hangs in equilibrium. One string is inclined at an angle of 30° to the horizontal and the tension in the string is 40 N. The second string is inclined at an angle of 60° to the horizontal. Calculate in newtons

(a) the weight of the particle,

(b) the magnitude of the tension in the second string.

**16** A particle $P$ is place on the inner surface of a fixed hollow sphere of centre $O$. Given that the coefficient of friction is 0.5 and that the particle rests in limiting equilibrium find the tangent of the angle between the downward vertical and $OP$.

**17** A particle of mass $m$ is in equilibrium on a rough plane inclined at an angle $\alpha$ to the horizontal. When a force of magnitude $mg$, in the sense up a line of greatest slope, is applied to the particle the latter is about to move up the plane. When a force of magnitude $\frac{mg}{2}$ in the sense down a line of greatest slope, is applied to the particle it is just about to move down the plane. Find $\sin \alpha$ and the coefficient of friction.

**18**

The diagram shows a circus artiste walking across a tightrope. The rope is tied at each end to a vertical pole and these poles are held in position by wires attached to their ends. The other ends of the wires are fixed to the ground and both wires are inclined at an angle of 60° to the horizontal. When the artiste, whose mass is 70 kg, is midway across the tightrope the tightrope is inclined at an angle of 80° to both poles. Find
(i) the tension in the tightrope,
(ii) the tensions in the supporting wires,
(iii) the thrust exerted at the top end of each pole.

**19** Two particles of the same mass are connected by an inextensible string. One particle lies on a rough plane inclined at an angle θ to the horizontal and the other hangs freely. The string connecting them passes over a smooth pulley which is above the particles and which separates the string into a part parallel to the inclined plane and a vertical part. Show that the system will move when released from rest if the coefficient of friction between the plane and the particle is less than sec θ – tan θ.

**20**

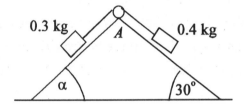

The diagram shows two particles, of mass 0.3 kg and 0.4 kg in equilibrium on two smooth inclined planes intersecting at a point $A$. They are joined by a light string passing over a small smooth pulley at $A$ and the particles and $A$ are in the same vertical plane. Find
(i) the tension in string,
(ii) the angle α.

**21**

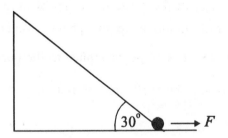

The diagram shows a heavy particle of mass 0.3 kg in equilibrium on a smooth horizontal plane. The particle is attached to a fixed point by a light string inclined at an angle of 30° to the horizontal. A horizontal force of magnitude $F$ is applied as shown.

Given that $F = 4$ N find
(i) the tension in the string,
(ii) the normal reaction of the plane on the sphere.
Discuss the behaviour of the reaction as $F$ is increased.

**22**

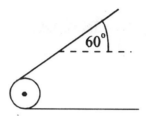

The diagram shows a rope passing over a fixed pulley; the directions at which the rope leaves the pulley are as shown. Assuming that the tensions in both parts of the rope are the same, and that the total force on the pulley is 3 kN, find the tension in the rope.

**23**

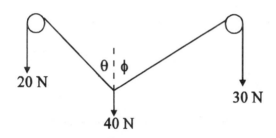

The diagram shows two strings attached to a particle of weight 40 N, the strings pass over two smooth pulleys at the same horizontal level and carry particles of weight 20 N and 30 N respectively at their ends. Given that the particles are in equilibrium verify that $\theta = 46.6°$, $\phi = 29°$.

This kind of apparatus forms the basis for verifying the " triangle law".

It is possible to calculate the values of the angles corresponding to any set of weights but this requires more trigonometry than you yet know. You can get over this however either by using scale drawing or using weights so that the strings at the point of intersection are perpendicular. (This will be true if the square of the weight at the point of intersection is the sum of the squares of the other weights.) You can then find the angles by resolving along the strings. If you can set up experiments with these weights you can compare your calculated values for the angles with the measured ones.

# Chapter 3

# Parallel forces acting on bodies

After working through this chapter you should be able to
- find the moment of a force about a point,
- solve problems of the equilibrium of bodies acted on by parallel forces.

## 3.1 Moment of a force

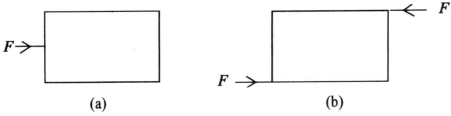

(a)　　　　　　　　　　(b)

If you push a book along a table by applying a force perpendicular to, and at the middle point of, one edge, as shown in diagram (a) above the book will move roughly in a straight line. On the other hand if you push the book by applying forces of the same magnitude, but of opposite directions, along two parallel edges as in diagram (b) above then the book will tend to rotate about its centre. If the book were pushed by applying a force at an arbitrary point of an edge then the motion would be a mixture of rotation and translation. Therefore, as the point of application moves, the effect of the force varies and in some circumstances there will be a turning effect. The moment of a force measures the tendency of a force to produce a rotation.

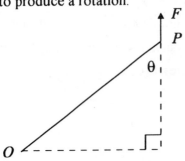

## Parallel forces acting on bodies

**Definition**

The magnitude of the moment, about a point $O$, of a force **F** acting through the point $P$ is the product of the magnitude of the force and the perpendicular distance from $O$ to the line of action of the force (this is the line through $P$ in the direction of the force).

If the force is measured in newtons and the distance in metres then the unit of moment is the newton metre, abbreviated to Nm.

The moment about $O$ is said to be "clockwise" or "anticlockwise" depending on whether the force is in the sense which would produce a clockwise or anticlockwise rotation about $O$. Moments of different forces have often to be added together and the convention used is that anticlockwise moments are positive and clockwise ones negative.

In the above diagram the perpendicular distance from $O$ to the line of action is $OP \sin \theta$ so the moment is of magnitude $F \times OP \sin \theta$. For the sense shown in the diagram the moment is positive.

If the line of action of the force passes through $O$ then the moment will be zero.

## Example 3.1

Find the moments of the following forces about $O$.

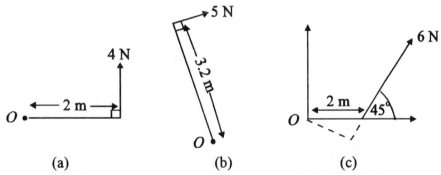

(a)   (b)   (c)

(a) The perpendicular distance of $O$ from the line of action is 2 m, the sense of rotation is anti clockwise so the moment is 8 Nm.

(b) The perpendicular distance of $O$ from the line of action is 3.2 m, the sense of rotation is clockwise so the moment is $-16$ Nm.

(c) The perpendicular distance of $O$ from the line of action is $2 \cos 45° \text{ m} = \sqrt{2}$ m, the sense of rotation is anti-clockwise so the moment is $6\sqrt{2}$ Nm.

It is possible to prove that the moment of a force about a point is the sum of the moments, about that point, of the components of the force in any pair of directions. This often gives a much easier way of working out the moment of a force than trying to find the perpendicular distance from the line of action.

## Example 3.2

Find the moment about $O$ of the force with $x$- and $y$- components $X$ N, $Y$ N and acting at the point $(a \text{ m}, b \text{ m})$ when

(a) $X = 3, Y = 4; a = 2, b = 5$    (b) $X = -3, Y = 4, a = -4, b = 5,$    (c) $X = 2, Y = -4, a = 2, b = 3;$    (d) $X = 3, Y = -4, a = 2, b = -5$.

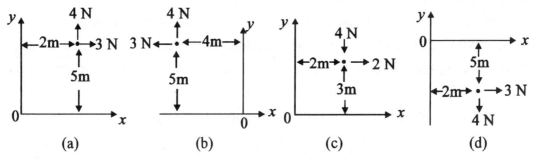

(a) The $y$-component has an anti clockwise moment of $4 \times 2$ Nm and the $x$-component has a clockwise moment of $3 \times 5$ Nm. The total moment is therefore $(8-15)$ Nm $= -7$ Nm.

(b) The $y$-component has a clockwise moment of $4 \times 4$ Nm and the $x$-component has an anti-clockwise moment of $3 \times 5$ Nm. The total moment is therefore $(-16+15)$ Nm $= -1$ Nm.

(c) The $y$-component has a clockwise moment of $4 \times 2$ Nm and the $x$-component has a clockwise moment of $2 \times 3$ Nm. The total moment is therefore $(-8 - 6)$ Nm $= -14$Nm.

(d) The $y$-component has a clockwise moment of $4 \times 2$ Nm and the $x$-component has an anti clockwise moment of $3 \times 5$ Nm. The total moment is therefore $(-8 +15)$ Nm $= 7$ Nm.

## Example 3.3

Find the moment about the point $Q$ $(x_0 \text{ m}, y_0 \text{ m})$ of the force with $x$ and $y$ components $X$ N and $Y$ N and acting at the point $P$ $(a \text{ m}, b \text{ m})$ when

(a) $X = 3, Y = 4; a = 2, b = 5, x_0 = 1, y_0 = 2$

(b) $X = -3, Y = 4; a = -4, b = 5, x_0 = 1, y_0 = 3$.

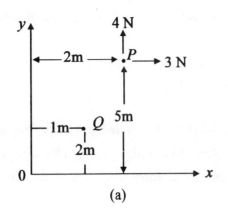
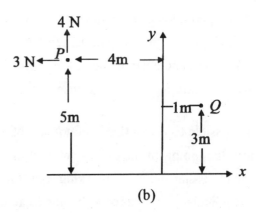

(a) The y-component has an anti-clockwise moment of 4 × 1 Nm and the x-component has a clockwise moment of 3 × 3 Nm. The total moment is therefore (4 − 9) Nm = −5Nm.
(b) The y-component has a clockwise moment of 4 × 5Nm and the x-component has an anti-clockwise moment of 3 × 2 Nm. The total moment is therefore (−20 + 6) Nm = −14Nm.

**Example 3.4**
The line of action of a force of magnitude 6 N passes through the point (1,2) and the direction of the force makes an angle of 60° to the x-direction. Find the moment of the force about the point (4,6).

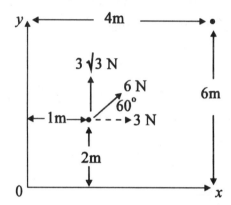

The force has components 3 N and $3\sqrt{3}$ N along the x- and y-axes, as shown in the diagram. The y-component has a clockwise moment of $3\sqrt{3} \times 3$ Nm whilst the x-component has an anti-clockwise moment of magnitude 3 × 4 Nm. The total moment is therefore $(12 - 9\sqrt{3})$ Nm = −3.59 Nm.

It is possible to derive a formula for the moment about the point $(x_0, y_0)$ of the force with components $(X, Y)$ acting at the point $(x, y)$. The moment is $(x - x_0)Y - (y - y_0)X$. It is not worth trying to remember this (though you should try and derive it) but it can be useful in checking.

**Moment of several forces**
The moment about a point of several forces (not necessarily acting through one point) is defined to be the algebraic sum of the moments about the point of the individual forces.

## Example 3.5
Find the moment about $O$ of the following system of forces. The forces are all perpendicular to the dotted lines.

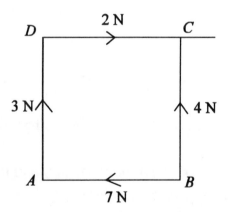

The force acting at $A$ has an anti-clockwise moment of 6.3 Nm, that at $B$ has a clockwise moment of 5.1 Nm and the force at $C$ has an anti-clockwise moment of 3.2 Nm.
The total moment is therefore $6.3 + 3.2 - 5.1$ Nm $= 4.4$ Nm.

## Example 3.6
Forces of magnitude 7 N, 4 N, 2 N, and 3 N act in the senses shown in the diagram along the sides $BA$, $BC$, $DC$ and $AD$ respectively of the square $ABCD$ of side 3 m. Find the moment about $A$ and about $C$ of the system of forces.

The forces whose lines of action pass through $A$ will have no moment about $A$. The force along $BC$ will have a anticlockwise moment of magnitude $4 \times 3$ Nm $= 12$ Nm, and the force along $CD$ will have a clockwise moment of $2 \times 3$ Nm $= 6$ Nm. The total moment is therefore $(12 - 6)$ Nm $= 6$ Nm.
The forces whose lines of action pass through $C$ will have no moment about $C$. The force along $BA$ will have a clockwise moment of magnitude $7 \times 3$ Nm $= 21$ Nm, and the force along $AD$ will have a clockwise moment of $3 \times 3$ Nm $= 9$ Nm.
The total moment is therefore $(-21-9)$ Nm $= -30$ Nm.

## Exercises 3.1

In questions 1 to 4 forces are shown acting at various points on a straight line and the distances are measured in metres. Unless otherwise indicated the forces are perpendicular to the line. Find the moments of the systems about the points A and B.

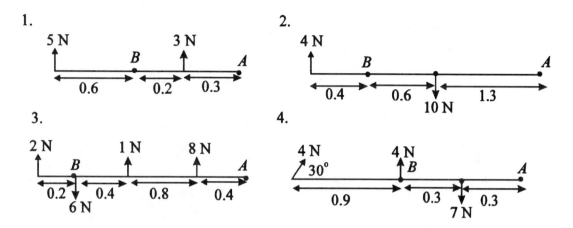

In questions 5 to 7 find the moments, about $O$, of the systems of forces shown. All forces are perpendicular to the dotted lines.

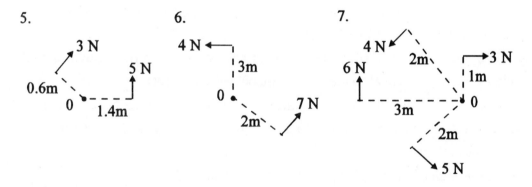

In the following three questions the moments of the systems of forces given are to be found about the point with coordinates, in metres $(a, b)$. The unit of force is the newton and the unit of distance is the metre.

**8** Force with components $(2,3)$ acting at $(1,1)$ and a force with components $(5,4)$ acting at $(3,-1)$, $a = 0$, $b = 0$.

**9** Force with components $(5,-3)$ acting at $(-4,1)$ and a force with components $(-2,-1)$ acting at $(5,-1)$, $a = 2$, $b = 1$.

**10** Force with components $(3,1)$ acting at $(2,-3)$ and a force with components $(-5,2)$ acting at $(-6,-1)$, $a = -3$, $b = 2$.

In the following questions find the moments of the systems shown about the points A and B.

*Parallel forces acting on bodies*

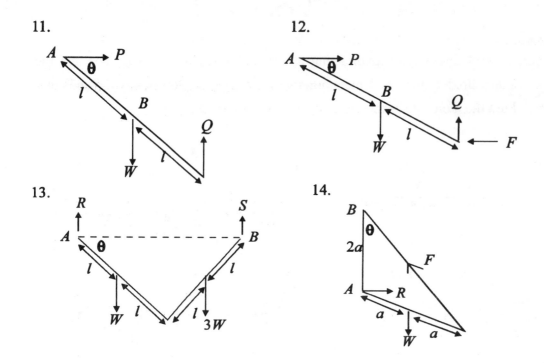

## 3.2 Equilibrium of a body acted on by parallel forces.

The conditions for the equilibrium of a body when a system of parallel forces acts are
(a) The sum of the components of all the forces is zero,
(b) The sum of the moments of all the forces about any point is zero (or, equivalently, the total clockwise moment is equal to the total anti-clockwise moment).
(Since the forces are all parallel the components will all be parallel to one direction and therefore (a) effectively simplifies to requiring the component in one direction to be zero.)
You can replace both of these by
(c) The moment of all the forces about any two points is zero.
In most cases it is simpler to use (a) and (b) but you should always remember that you can only get two conditions per body.
(If, for example, you got three equations by using (a) and (c) you would find that the third equation could have been obtained from the other two. The third equation therefore would not have given any more information).
The important thing, again, is to draw clear force diagrams and then apply the conditions. If two bodies are in contact you may have to consider them separately but you must remember Newton's third law in making the force diagrams.
The choice of the point about which to take moments is again yours and you should try to choose it so that one unknown force is eliminated by taking moments about a point on the line of action of that force.

## Parallel forces acting on bodies

There are no new modelling assumptions made but you should remind yourself of the meanings of the phrases in the Glossary in Chapter 2. The only slightly different feature is that, for bodies that are not light, the force of gravity will act. This acts through the centre of gravity. You will not be expected, in general, to know the position of the centre of gravity but you should know that the centre of gravity of a uniform rod is at its midpoint.

**Example 3.7**

The diagram shows a light rod $AB$, of length 0.8 m, pivoted at a point $C$ a distance 0.5 m from $A$. Find the mass that has to be placed at $B$, so that the rod will stay horizontal with a mass of 0.3 kg at $A$.

The force diagram is shown in the right hand diagram above.

Since there is a pivot at $C$ it is better to take moments about it so that the reaction at the pivot will not enter into the moment equation. If the mass at $B$ is denoted by $m$ kg then equating the moments gives

$$0.3 \times 0.5 = m \times 0.3,$$
so $\quad m = 0.5.$

**Example 3.8**

A light beam $AB$ of length 8 m has a load weight of 90 N attached to its midpoint $O$. The beam rests horizontally on two smooth pegs $C$ and $D$ with $AC = 2$ m, $DB = 3$ m and with loads of weight 12 N and 20 N attached at the ends $A$ and $B$ respectively. Find the reactions at the pegs.

The force diagram is:

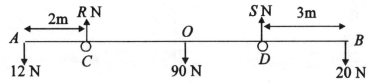

The pegs are smooth so their reactions are vertically upwards. Resolving vertically gives

$$R + S = 122.$$

If moments are taken about $C$ then the reaction $R$ will not occur in the equation. Therefore

$$90 \times 2 + 20 \times 6 = 12 \times 2 + S \times 3,$$

giving $S = 92$ and hence $R = 30$.

## Example 3.9

Find, for the previous example the greatest weight that can be placed at $B$ without disturbing equilibrium.

Obviously for a sufficiently large weight at $B$ the rod will start turning about $D$ and when this happens $R = 0$. If the weight is denoted by $W$ then resolving vertically gives
$$R + S = 102 + W.$$
The moment equation now becomes
$$90 \times 2 + W \times 6 = 12 \times 2 + S \times 3.$$
If $R = 0$ then, from the first equation, $S = 102 + W$, substituting this in the second equation gives $W = 50$.

In finding $W$ a guess was made about what would happen. It is possible to avoid the guess with very little extra work. Solving the moment equation for $S$ gives $S = 52 + 2W$, and this gives $R = 50 - W$. This shows that $R$ would be negative for $W > 50$, so contact would have been lost.

## Example 3.10

Two light rods $AB$ and $BC$ are rigidly connected at $B$ so that they are at right angles to each other. $AB$ is of length 3 m, $BC$ is of length 4 m, and weights $W$ and $3W$ are attached to $A$ and $C$ respectively. The configuration is suspended in equilibrium by a light string attached to $B$. Find the inclination of $AB$ to the horizontal.

The force diagram is :

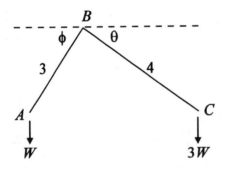

Taking moments about $B$ gives
$$3W \cos \phi = 12 \, W \cos \theta.$$
Since $\theta + \phi = 90°$, $\cos \phi = \sin \theta$ and therefore $\tan \theta = 4$ and $\theta = 76°$.

## Parallel forces acting on bodies

**Exercises 3.2**

Questions 1 to 4 involve a light rod, simply supported at points $C$ and $D$, and acted upon by the forces shown. The distances are measured in metres. Find the forces acting at $C$ and $D$.

1.

```
A          C         D    B
↓   0.6    ○   0.8   ○ 0.2 ↓
7 N                         9 N
```

2.

```
A        C           D    B
↓ 0.2    ○   0.4  ↓  0.3 ○ 0.1 ↓
8 N              7 N           2 N
```

3.

```
  7 N
  ↑
A      C            D       B
|  0.3 ○ 0.2 ↓ 0.3 ○  0.5  ↓
              22 N          1 N
```

4.

```
A    C                  D    B
     ○ 0.2 ↓ 0.3 ↓ 0.6       ○
        4 N    5 N
```

**5** A uniform rod $AB$ of length 4 m and mass 4 kg is pivoted about the point $C$ where $AC = 1$ m. Find the mass of the particle which must be attached at $A$ so that equilibrium is possible with the rod horizontal.

**6** A uniform rod $AB$ of length 6 m and mass 8 kg has a mass 12 kg attached at $A$ and a mass 16 kg attached at $B$. Find the position of the point about which the rod can be balanced in a horizontal position.

**7** A see-saw is made of a heavy plank of mass 30 kg and length 5 m and is pivoted at its midpoint. Two children of mass 25 kg and 35 kg can sit, one at each end of the seesaw with the latter horizontal.
Find the distance of the centre of gravity of the plank from the child of mass 25 kg. State two modelling assumptions that you make.

**8** A light rod $AB$ of length 2 m is simply supported at points $C$ and $D$, where $AC = 0.6$ m, $BD = 0.3$ m. A downward force of magnitude 60 N is applied at the midpoint of $AB$. Find the forces acting at the pegs. Find also the least force acting downwards at $A$ which will disturb equilibrium.

**9** A light square lamina $ABCD$ is free to turn about its centre in a vertical plane. A particle of mass $m$ is attached at $A$. Find the mass of the particle that has to be attached at $B$ so that the square can be in equilibrium, with $AB$ inclined at an angle of $30°$ above the horizontal and $A$ lower than $B$.

## 3.3 Resultant of a number of parallel forces

The idea of resultant can be extended to a number of parallel forces. If the sum of the components of all the forces is not zero the resultant is a force whose component is the sum of the components of all the separate forces. It acts through a point such that the moment of the resultant about any point is the sum of the moments of all the other forces about that point.

The point through which the resultant acts is such that the sum of the clockwise moments about that point is equal to the sum of the anti-clockwise moments about that point.

This definition is only valid when the sum of the components is not zero. If the sum of the components is zero then the resultant is not a force but a couple. A couple is effectively two parallel forces, acting through different points, not on the same line, of equal magnitude but opposite directions.

You will not be expected to know anything about couples.

If you know the resultant of a system of parallel forces then you can obtain equilibrium by adding an equal and opposite force to the resultant.

### Example 3.11

Find the resultant of the forces shown below.

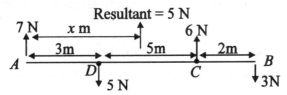

The component of the resultant up the page is $(7 + 6 - 5 - 3)$ N $= 5$ N. Taking moments about a point between $D$ and $C$ and at a distance $x$ m from $A$, the anti-clockwise moments is $5(x - 3) + 6(8 - x)$, and the clockwise moment is $7x + 3(10 - x)$. Therefore $5(x-3) + 6(8 - x) = 7x + 3(10 - x)$ i.e $5x = 3$. The resultant is therefore a force of magnitude 5 N acting at a distance of 0.6 m from $A$.

### Exercises 3.3

Find the resultant of the forces shown, the distances are all measured in metres.

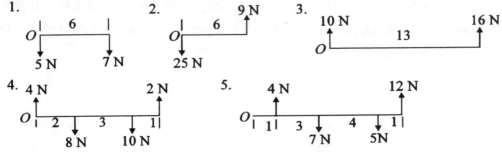

## Miscellaneous Exercises 3

**1**

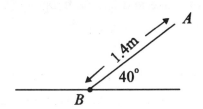

The diagram shows a light rigid rod $AB$ of length 1.4 m inclined at an angle of $40°$ to the horizontal with the end $B$ fixed to horizontal ground.

(a) Find the moment about $B$ of a downward vertical force of magnitude 120 N applied at $A$.

(b) Find the magnitude and direction of the force of least magnitude that can be applied at $A$ and whose moment about $B$ is equal to that found in (a).

**2** A uniform beam $AB$, of length 8 m and weight 200 N, rests horizontally on two smooth supports at points $C$ and $D$, where $AC = 1$ m and $AD = 6$ m. Loads of 100 N and 400 N are attached to the beam at points $E$ and $F$, where $AE = 2$ m and $AF = 5$ m. Find the reactions at the supports.

**3** A uniform plank $AB$, of length 4 m and weight 200 N, rests horizontally on two smooth supports at points $C$ and $D$, where $AC = 0.5$ m and $AD = 3.2$ m, with a load $W$ N attached at $B$.

(a) Given that $W = 84$, find the reactions at the support.

(b) Find the greatest value of $W$ for which equilibrium is possible.

**4** A non-uniform beam $AB$, of length 10 m and weight 100 N, rests horizontally on two smooth supports at points $C$ and $D$, where $AC = 3$ m and $AD = 8$ m. When a weight of 200 N is suspended from the midpoint of $AB$ the magnitude of the reaction at $C$ is twice the magnitude of the reaction at $D$. Find the distance of the centre of gravity of the beam from $A$.

**5**

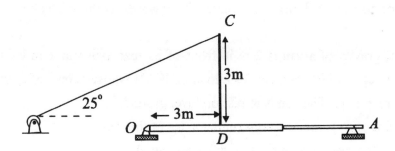

The diagram shows a flagpole $OA$ with a rigid rod $CD$ attached to it. A rope is attached to $C$ and passes over a winch which is used to wind the rope and lift the flagpole. Given that the tension in the rope is 3 kN find the moment of this tension about $O$.

**6** A motorcycle has mass 200 kg, the points of contact of the wheels with the road are 1.5 m apart and the line of action of the weight is at the same distance from both wheels. The rider is of mass 75 kg and the line of action of his weight intersects the road at a distance of 0.9 m from the front wheel. Find the reaction of the road on the two wheels.

**7** A woman of mass 75 kg crosses a garden stream by using, as a bridge, a plank of length 3 m and mass 150 kg. Assume that
- the plank can be modelled as a rod with its centre of gravity at its centre,
- the woman can be modelled as a particle,
- the reactions of the banks of the stream act at the ends of the plank.

Find the reactions of the banks on the plank given that the woman is 0.5 m along the plank.

What further information would you need to find the reactions when the woman is pushing a wheelbarrow along the plank?

**8**

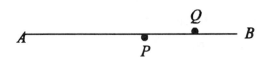

The diagram shows a light rod $AB$ of length $8a$ resting horizontally between two smooth pegs $P$ and $Q$, where $AP = 5a$ and $QB = a$. A particle of weight $4W$ is attached at the midpoint of the rod. Find the reactions of the pegs.

The greatest force that the peg $Q$ can sustain is $10W$. Find the greatest magnitude of the force that can be applied at $B$, (a) downwards, (b) upwards, without breaking equilibrium.

**9** The centre of gravity of a van is 2 m in front of its rear axle and 1 m behind its front axle. The van weighs 15 kN and carries a load of $W$ kN whose centre of gravity is 0.5 m in front of the rear axle. The van is at rest on level ground.
(a) Find, assuming $W = 3$, the reactions on the axles of the van.
(b) Find $W$ when the reactions on the two axles are equal.

**10**

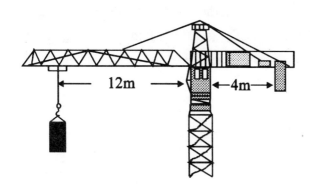

The diagram shows a tower crane carrying a load of mass 25 tonne. Find, assuming that the mass of the cabin and the structure may be neglected, the mass of the counterweight necessary to achieve equilibrium. Find also, assuming the load to be increased to 80 tonne, the new position of the load.

**11** A light rod $ABGCD$ with a load applied at $G$ rests horizontally on two smooth supports at $B$ and $C$. The lengths $AB$, $BC$ and $CD$ are 1.2 m, 2.4 m and 1.8 m respectively. The rod just starts to tilt when a load of 150 N is attached at $A$ or when a load of 40 N is applied at $D$. Find

(a) the load at $G$,

(b) the length $AG$,

(c) the reactions on the rod when loads of 150 N and 60 N are simultaneously applied at $A$ and $D$.

**12** A light beam $AB$ of length $6a$ rests horizontally on smooth pegs at its points of trisection. A heavy particle is placed at a point $P$ of the beam. When a mass of 1 kg is placed at $A$, the beam just tilts. When a mass of 1 kg is placed at $B$ the reactions at both supports are equal. Find the mass of the particle and the distance $AP$.

**13**

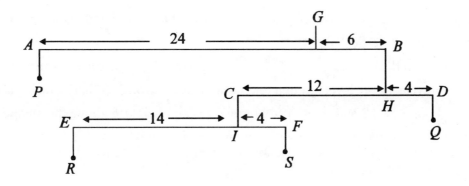

The diagram shows three light rods $AB$, $CD$ and $EF$ with particles $P$, $Q$, $R$, $S$ suspended from $A$, $D$, $E$ and $F$ respectively. Light strings $BH$ and $CI$, as shown, connect the three rods. The system is suspended in equilibrium from the point $G$ on $AB$ and, in equilibrium, the rods are horizontal. Given that the weight of $Q$ is 13.5 N find the weights of the other particles. All the distances shown are in centimetres.

**14**

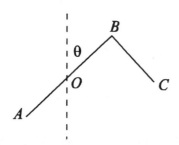

The diagram shows a light rod $AB$ of length $4a$ rigidly joined at $B$ to a light rod $BC$ of length $2a$ so that the rods are perpendicular to each other and in the same vertical plane. The centre $O$ of $AB$ is fixed and the rods can rotate freely about $O$ in a vertical plane. A particle of mass $4m$ is attached at $A$ and a particle of mass $m$ is attached at $C$. The system rests in equilibrium with $AB$ inclined at an acute angle $\theta$ to the vertical as shown. By taking moments about $O$ find the value of $\theta$.

**15** A light beam $AD$ of length $6a$ rests horizontally on smooth pegs at its points of trisection $B$ and $C$. A particle of weight $W$ is placed at its midpoint. A man of weight $4W$, in order that he can stand at $A$, places a counter weight $W'$ at a point $P$, between $C$ and $D$, and at a distance $x$ from $D$. Find the least value of $W'$.

Find also the maximum value of $W'$ so that the beam will still be in equilibrium when the man is not standing on it.

**16** A motor cycle which is too long to go on a scales is weighed by placing first one wheel and then the other wheel on the scales. The platform of the scales is above the level of the ground.

During the weighings the lower wheel is in contact with the ground and clamped so that the motor cycle remains vertical.

The distance between the centres $A$ and $B$ of the wheels (which are of equal diameter) is $c$. When the motorcycle is on level ground the weight acts through a line which passes through the mid point of $AB$ and the centre of gravity is at a height $h$ above this midpoint. The inclination of $AB$ to the horizontal when either wheel is on the scales is $\theta$. Show that the sum of the two weighings is less than the actual weight $W$ by $\dfrac{2Wh}{c}\tan\theta$.

# Chapter 4

# Kinematics of Rectilinear Motion

After working through this chapter you should
- understand what is meant, in rectilinear motion, by displacement, velocity and acceleration,
- be able, given one of displacement, velocity or acceleration as a function of time, to find the other two,
- be able to derive the "constant acceleration formulae" and use them to solve problems involving motion under constant acceleration,
- be able to solve simple problems of vertical motion under gravity.

## 4.1 Basic definitions

When trying to solve any problem involving the rectilinear motion (that is motion in a straight line) of a particle, it is essential at the outset to choose a particular direction to be the positive direction and to refer everything to this direction. The choice of reference direction does not matter; the important thing is to stick to the same reference direction throughout a given problem. Failing to do this is the greatest source of error in problems, particularly in setting up equations of motion. For simplicity in what follows motion will be assumed to be along the $x$-axis and the positive direction to be that of increasing $x$.

**Displacement**

The position of a particle at any time is determined by its $x$-coordinate and in kinematics this coordinate is referred to as the displacement of the particle from the origin. Of course, the displacement $x$ can be positive or negative depending on whether the particle lies to the right or the left of the origin. The distance of the particle from $O$ is $|x|$. By definition, the displacement of a particle specifies its position uniquely, whereas the distance from the origin does not do this because it does not identify the side of the origin on which the particle lies.

The usual unit of measurement in the S I system is the metre, (abbreviation m) though for small distances the centimetre (cm) is used and, for large distances, the kilometre (km) is used.

In principle it is possible by measuring the displacement of a particle, to express $x$ as a function of $t$ and diagrams (a), (b), (c) and (d) below show the behaviour of $x$ as $t$ varies for the four cases

(a) $x = 2t$,  (b) $x = \frac{1}{4}(t-4)^2$,  (c) $x = 4 - t^2$,  (d) $x = \frac{1}{4}(t^3 - t^2) + 1$.

You will see that, in (c), $x$ becomes negative for $t > 2$.

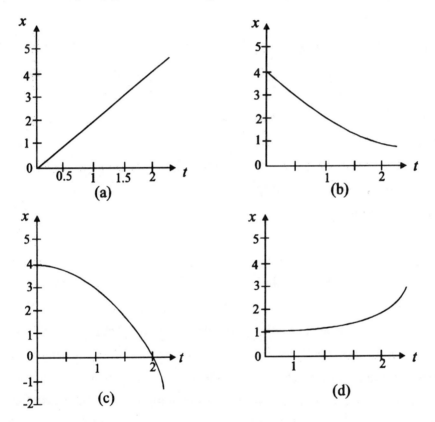

It is worth looking more closely at diagram (a) where $x = 2t$. For simplicity it will be assumed that the displacement is measured in metres and time in seconds. The difference between the displacements at times $T$ and $T + t'$, is $2t'$ for all values of $T$, i.e. the same distance is travelled in the same time interval. You can also see this from the diagram. Therefore in any interval of 1 second the particle travels 2 m and you would intuitively interpret this as saying that the particle has a speed of 2 ms$^{-1}$. It is not possible to give a simple interpretation in the other diagrams since you can see that equal distances are not covered in equal times. To cover these cases it is necessary to have a more sophisticated approach as follows.

## Velocity

The velocity $v$ in the positive direction is defined to be the rate of change of the displacement with respect to time, that is,

$$v = \frac{dx}{dt} = \dot{x}.$$

Superscript dots are often used to imply differentiations with respect to time $t$.

The unit of velocity commonly used in the S.I. system is the metre per second (ms$^{-1}$), and for velocities of relatively large magnitude the kilometre per hour (kmh$^{-1}$) is used.

If you have not yet come across differentiation in your course then there is a simple graphical way of defining the rate of change. This is that the rate of change of $x$ with $t$ at any time is the slope of the graph of $x$ against $t$ at that particular time. The diagrams below show the velocities corresponding to the four graphs of displacement above.

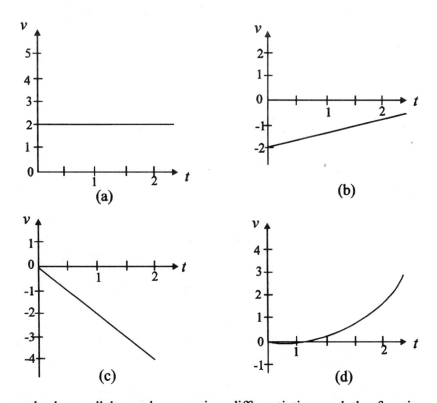

The graphs have all been drawn using differentiation and the functions defined are 2, $\frac{1}{2}(t-4)$, $-2t$ and $\frac{1}{4}(3t^2 - 2t)$. If you have not yet met differentiation you can check the graphs by drawing the tangents to the graphs of $x$ as a function of $t$ and see that the slopes are the values of $v$ in the second set of graphs. In diagrams (b) and (c) the velocity is negative, showing, as you can check from the corresponding diagrams for $x$, that $x$ decreases with time.

The velocity can again be either positive or negative and there is no direct dependence between the signs of $x$ and $v$. For example, $x = 1 - t^2$ is positive for $0 < t < 1$, whereas the velocity, which is $-2t$, is negative for $0 < t < 1$. Also in diagram (b) above the velocity is negative though $x$ is positive. All that is implied by a negative velocity is that the motion is in the opposite direction to the reference one. The speed is the magnitude of the velocity, that is, speed is equal to $|v|$. Velocity and speed as defined can vary with time but in diagram (a) above the velocity is constant, or uniform. In many practical situations the speed is estimated by dividing the total distance by the total time, this is however only a very rough estimate and should not be used to calculate the velocity at a particular time. The ratio distance/total time defines the average speed.

**Acceleration**

In diagrams (b) and (c) above the slopes are constant, and therefore the rate of change of velocity can be worked out easily. This is not true for diagram (d) and in order to cover this a further important idea has to be introduced namely that of acceleration.

The acceleration $a$ in the positive direction is defined as the rate of change with respect to time of the velocity in the positive direction, that is,

$$a = \frac{dv}{dt} = \frac{d^2x}{dt^2} = \ddot{x}$$

Again this can be positive or negative. The units of acceleration are metres per second$^2$ (ms$^{-2}$) and kilometres per hour$^2$ (kmh$^{-2}$). When there is no possibility of misunderstanding, the phrase 'in the positive direction' is omitted after velocity and acceleration, but it should be remembered that it is always implied. If a particle is said to be moving with retardation $r$, then conventionally this means that the acceleration in the reference direction is equal to $-r$. A retardation is sometimes called a deceleration and a particle is sometimes said to decelerate.

The acceleration is the slope of the graph of $v$ against $t$. For diagram (a) above the acceleration is zero and for diagrams (b), (c) and (d) the graphs of acceleration against time are given below.

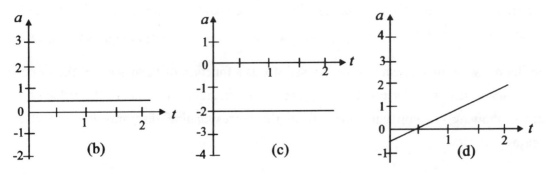

The accelerations in the three cases are $\frac{1}{2}$, $-2$ and $\frac{1}{2}(3t - 1)$ and again you can check them by finding the slopes of the previous sets of curves. In the first two cases the acceleration is constant, the second case corresponding to a retardation.

In most practical situations the acceleration $a$ will be given and may depend on all three of $t$, $x$ and $v$ and $x$ or $v$ have to be found. Finding $x$ from $a$ requires the solution of a differential equation; some simple examples of the general type of problem are discussed in 4.4 but it is possible to make considerable progress with the case of constant acceleration without elaborate calculation.

## 4.2 Constant acceleration

For a particle moving under constant (or uniform) acceleration $a$, with initial velocity (i.e. its velocity when $t = 0$) $u$, it is possible to give formulae for its displacement $s$ from its initial position and its velocity $v$ at time $t$. These are

$$v = u + at, \quad \ldots\ldots (1)$$

$$s = ut + \frac{1}{2}at^2, \quad \ldots\ldots (2)$$

$$v^2 = u^2 + 2as, \quad \ldots\ldots (3)$$

$$s = \frac{1}{2}(u + v)t. \quad \ldots\ldots (4)$$

If the initial conditions were given at $t = T$ and not at $t = 0$ then $t$ would have to be replaced by $t - T$. The derivation of these equations is given in 4.3.

Equations 1 to 4 are sufficient to enable all problems involving motion with constant acceleration to be solved, and they should be committed to memory.

It is also very important to remember that they can only be used for <u>constant acceleration</u>.

Before describing the use of these equations in solving problems, it is useful to give a graphical interpretation of equation 1 which is very useful in solving particular types of problems. The behaviour of $v$, as defined by equation 1, with $t$ is shown below.

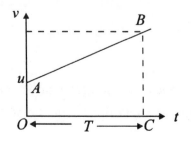

It is a straight line with gradient $a$. The area of the region between the $t$-axis and under the line, and between the lines $t = 0$ and $t = T$, is the area of the trapezium $OABC$ and this is $\frac{1}{2}(u + v)T$ which, by equation 4, is equal to $s$. This is a special case of the general result (shown in 4.3), that the area under the graph of $v$ against $t$ is equal to the distance covered.

Effectively, all the information supplied by equations 1, 2 and 4 is contained in the above diagram when the area $OABC$ is interpreted as $s$. Such a graph can be a useful, compact way of setting down the conditions in a given problem and is referred to as the velocity - time, or $v - t$, diagram.

**Problem solving**

The simplest kind of problem for motion under constant acceleration is the one in which $u$ and $a$ are given, and values of $s$ and $v$ are required for particular values of $t$. These problems are solved by substitution in equations 1 to 4. In solving problems it is important that all the quantities in equations 1 to 4 are evaluated in the same unit system. In the following examples, the S.I. unit system will be used so that the displacement from the initial position is denoted by $s$ m, the initial and final speeds are denoted by $u$ ms$^{-1}$ and $v$ ms$^{-1}$ respectively, the acceleration by $a$ ms$^{-2}$ and the time by $t$ s. This means that there are no units associated with $s$, $u$, $v$, $a$ and $t$ and they are, therefore, just numbers which satisfy equations 1 to 4.

**Example 4.1**

Initially, a particle $P$ moving with uniform acceleration 6 ms$^{-2}$ has a velocity at the point $O$ of 3 ms$^{-1}$. Find its velocity after 2 s and its displacement from $O$ after 4 s.

This is the simplest kind of problem, since both $u$ (= 3) and $a$ (= 6) are given. The velocity after 2 s is found by setting $t = 2$ in equation 1, giving
$$v = 3 + 12 = 15,$$
so the velocity is 15 ms$^{-1}$. The distance travelled in the first 4 s is found by substituting $t = 4$ into equation 2, giving
$$s = 3 \times 4 + 3 \times 16 = 60.$$
The displacement from $O$ is therefore 60 m.

A slightly harder class of problem arises when $u$ and $a$ are not given directly, but sufficient information is available to find them. In solving this kind of problem, the best method is to list the unknowns and then find them systematically by choosing whichever of equations 1 to 4 contains only one unknown. This equation then gives that unknown. First try to find $u$ and $t$. Then $s$ and $v$ can be found for all values of $t$ by using equations 1 to 4.

## Example 4.2

The velocity of a particle $P$ moving with a uniform acceleration of 3 ms$^{-2}$ increases from 2 ms$^{-1}$ to 8 ms$^{-1}$ as $P$ moves from $A$ to $B$. Find the distance between $A$ and $B$.
In this case $v = 8$, $u = 2$, $a = 3$, $s$ is required and this suggests using equation 3. Making the appropriate substitutions gives
$$64 = 4 + 6s,$$
so that $s = 10$. The points $A$ and $B$ are therefore 10 m apart.
This problem could also have been solved by substituting in equation 1 to find the total time, giving $t = 2$, and then using equation 4 to obtain the value of $s$.

## Example 4.3

A train starts from rest at a station and moves with constant acceleration. Twenty seconds later it is moving with speed of 72 kmh$^{-1}$ when it passes a signal box. Find the distance, in metres, between the station and the signal box.
In this question two units of length and two units of time are involved so the first thing is to decide on the units to be used and, since an answer is required in metres, it seems reasonable to use metres and seconds.
$$72 \text{ kmh}^{-1} = \frac{72 \times 1000}{3600} \text{ ms}^{-1} = 20 \text{ ms}^{-1}.$$
In this question $t$ (=20), $u$ (=0), $v$ (=20) are given so $a$ can be found from equation 1 and
$$20 = a20,$$
so that $a = 1$. The distance can now be found from equation 3 i.e.
$$400 = 2 \times 1 \times s,$$
so that the distance from the station to the signal is 200 m.

## Example 4.4

The displacement from its original position of a particle moving with uniform acceleration is 6 m after 2 s and 20 m after 4 s. Find the displacement 6 s after the start of the motion.

Neither $a$ nor $u$ is given, but values of $s$ are given for two values of $t$ i.e. $s = 6$ when $t = 2$ and $s = 20$ when $t = 4$. Substituting into equation 2 for these two values gives
$$6 = 2u + 2a, \quad 20 = 4u + 8a.$$

Solving these simultaneously gives $a = 2$ and $u = 1$. The displacement after 6 s is found by substituting these values in equation 2, with $t = 6$, giving $s = 6 + 36 = 42$, and therefore the required displacement is 42 m.

**Example 4.5**
A boy moving up a hill on a skate-board experiences a retardation of magnitude 2 ms$^{-2}$. His speed at the bottom of the hill was 8 ms$^{-1}$, find how far up the hill he travels before coming to rest.
Since the boy experiences a retardation his acceleration in the sense up the hill is $-2$ ms$^{-2}$. In this case $u = 8$, $v = 0$ and $a = -2$ so $s$ can be found from equation 3 i.e.
$$0 = 64 - 2 \times 2s,$$
giving $s = 16$ so that the boy travels 16 m up the hill before coming to rest.

**Example 4.6**
A car, travelling at 20 ms$^{-1}$, has to be braked suddenly and skids a distance of 25 m before stopping. Find the acceleration, assuming that it is constant, and the time taken to stop.
In this case $v = 0$, $u = 20$ and $s = 25$ so equation 3 can be used to find $a$ i.e.
$$0 = 400 + 2 \times 25\, a,$$
giving $a = -8$, so that the acceleration is $-8$ ms$^{-2}$. This is a retardation and is to be expected since the car is slowing down. The time can now be found from equation 1 which gives
$$0 = 20 - 8t,$$
giving the time to be 2.5 s.

Probably the most complicated problems involving constant acceleration are those in which the acceleration is constant for a particular period but then switches to another constant value for a different period. This kind of problem can occur, for example, in the motion of a train which accelerates from rest to a steady speed, keeps that steady speed for a while, and then retards to come to rest. In such problems, equations 1 to 4 have to be applied systematically for each period, and the information given in a question used to find all the unknowns. It is in these problems, where the given information can be complicated, that the $v$ - $t$ diagram discussed is most useful. The given information can be displayed compactly on a diagram and elementary geometry used to complete the question. This approach is particularly useful for 'rest-to-rest' problems. The graph of $v$ against $t$ will be a series of straight lines, with the parts with constant velocity being segments parallel to the $t$- axis.

## Kinematics of Rectilinear Motion

### Example 4.7

Starting from rest, a train moves with uniform acceleration and reaches a maximum speed of 20 ms$^{-1}$ in 50 s. It runs at this speed for 35 s and then comes to rest with uniform retardation in 40 s. Find the total distance travelled.

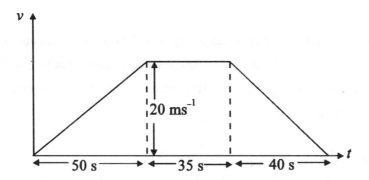

The diagram shows the information set out on a $v$ - $t$ diagram. The distance is the total area of the figure. The parallel sides of the trapezium are of lengths 125 and 35 respectively. The area is therefore $\frac{1}{2}(125 + 35) \times 20 = 1600$ so that the distance travelled is 1600 m.

### Example 4.8

Over a 100 m length, a runner accelerates uniformly from 6 ms$^{-1}$ to 10 ms$^{-1}$ and then maintains the latter speed over the remaining length. Given that the total time for the 100 m distance is 11 s, find the acceleration.

The $v$-$t$ diagram is shown below.

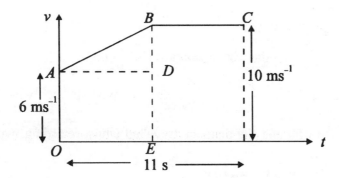

The total area under the lines $AB$ and $BC$ is 100, this area is also equal to
$$\frac{1}{2}(6 + 10)\,OE + 10(11 - OE)$$
and equating this expression to 100 gives $OE = 5$.

The gradient of $AB$ is the unknown acceleration $a$ ms$^{-2}$, so that
$$a = \frac{BD}{AD} = \frac{10-6}{5} = 0.8.$$
Therefore $a = 0.8$ so that the acceleration is 0.8 ms$^{-2}$.

**Example 4.9**

A train starting from rest moves with constant acceleration for 5 minutes. It then moves at a constant speed for 20 minutes. A constant retardation is then applied, whose magnitude is twice that of the acceleration, until the train comes to rest. Find, given that the train travels 4.5 km whilst accelerating

(i) the acceleration,

(ii) the total distance travelled.

The $v$-$t$ diagram is given below, where time is measured in seconds and velocity in ms$^{-1}$.

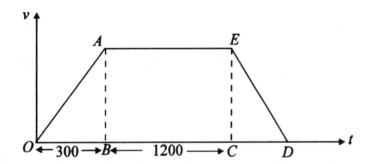

(i) The distance travelled whilst accelerating is the area under the line $OA$. The length $OB$ is 300 (corresponding to the time accelerating) so
$$4500 = \frac{1}{2} \times 300 \times AB,$$
giving $AB = 30$, i.e the constant speed is 30 ms$^{-1}$.

The slope of $OA$ is $\frac{AB}{OB} = \frac{30}{300} = \frac{1}{10}$, so that the acceleration is $\frac{1}{10}$ ms$^{-2}$.

(ii) The retardation is the slope of $DE = \frac{EC}{CD} = \frac{30}{CD}$, this is given to be twice the retardation so it is $\frac{1}{5}$ ms$^{-1}$ so $CD = 150$ and the distance travelled whilst retarding, which is the area under $DE$, is $\frac{1}{2} \times 30 \times 150$ m i.e. 2.25 km.

The length $BC$ is 1200 corresponding to the time the train is travelling at constant speed and therefore the distance travelled at constant speed is $30 \times 1200$ m $= 36$ km. The total distance travelled is therefore 4.5 km + 2.25 km + 36 km $= 42.75$ km.

## Kinematics of Rectilinear Motion

**Exercises 4.1**

Questions 1 to 11 refer to a particle moving on a straight line with constant acceleration $a$ ms$^{-2}$, so that at time $t$ s the velocity and the displacement of the particle from a fixed point $O$ are given by $v$ ms$^{-1}$ and $s$ m, the initial velocity of the particle being $u$ ms$^{-1}$. The acceleration, velocities and displacement are with respect to the same reference direction.

**1** $a = 4$, $v = 12$, $u = 4$, find $t$.
**2** $a = -5$, $u = 3$, $v = -12$; find $t$ and $s$.
**3** $u = 6$, $t = 2$, $s = 20$; find $v$.
**4** $u = -2$, $a = 3$, $t = 5$; find $s$.
**5** $a = 4$, $t = 4$, $s = 42$; find $u$.
**6** $v = 11$, $u = 7$, $t = 2$, find $a$.
**7** $v = 8$, $u = 6$, $s = 7$; find $a$.
**8** $s = 25$, $u = 4$, $v = 11$; find $t$.
**9** $s = 70$, $u = 4$, $t = 5$; find $a$.
**10** $s = 8$, $u = 9$, $a = -5$; find the possible values of $t$.
**11** $s = 25$ when $v = 13$ and $s = 52$ when $t = 4$; find the possible values of $u$ and $a$.

Questions 12 to 13 refer to particles $P$ and $Q$ free to move on adjacent parallel lines as shown in the diagram, $O$ and $O'$ are fixed points on the lines and $OO'$ is perpendicular to both lines.

**12** At time $t = 0$ s, $P$ passes through $O$ moving towards a point $B$ with a speed of 7 ms$^{-1}$ and thereafter moves with a constant acceleration of 2 ms$^{-2}$ directed towards $B$. At the same time, $Q$ passes through the point $B'$, which is directly opposite to $B$ and which is 114 m away from $O'$, with a velocity directed towards $O'$ of 3 ms$^{-1}$ and thereafter $Q$ has acceleration 1 ms$^{-2}$ in the direction $B'O'$. Find when $P$ and $Q$ are level with each other.

**13** At time $t = 0$ s, $P$ passes through $O$ with a speed of 4 ms$^{-1}$ and thereafter moves with a constant acceleration of 2 ms$^{-2}$. At time $t = 2$ s $Q$ passes through the point $C'$, 100 m away from $O'$, with a velocity directed towards $O'$ of 17 ms$^{-1}$. Thereafter $Q$ has an acceleration of 4 ms$^{-2}$ in the sense $C'O'$. (The distance between the tracks may be neglected). Find when $P$ and $Q$ are at a distance of 60 m apart.

**14**

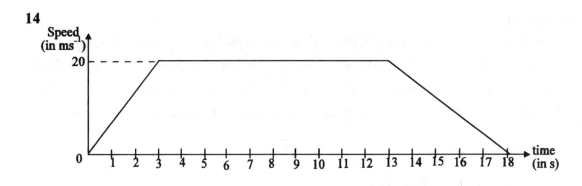

The diagram is a *v-t* diagram for a particle which moves with constant acceleration for 3 seconds, then moves with a constant velocity for 10 seconds and then moves with constant retardation until it comes to rest. Find
(i) the acceleration,
(ii) the retardation,
(iii) the total distance travelled.

**15** An underground train covers 576 m from rest to rest in 60 s. At first it has a constant acceleration of 0.5 ms$^{-2}$, then moves with constant speed and finally has a constant retardation of 1 ms$^{-2}$. Find the time taken for each stage of the journey.

**16** A train approaching a station travels two successive distances of 0.25 km in 10 s and 20 s respectively. Assuming the retardation to be uniform (i.e constant), find the further time taken before coming to rest.

**17** A train starting from rest is uniformly accelerated during the first 80 s of its journey in which it covers 600 m. It then runs at a constant speed until it is brought to rest in a distance of 750 m by applying a constant retardation. Find the maximum speed of the train and the magnitude of the retardation.

**18** A car starting from rest moves with constant acceleration of 1 ms$^{-2}$ for 10 seconds, it then moves at a constant speed for 1 km and then a constant retardation is applied to bring it to rest. The total distance travelled is 2 km. Find the maximum speed of the car and the total time taken.

**19** A car moves off from rest with a constant acceleration of 1.2 ms$^{-2}$ but after 12 s the driver sees an obstacle and is forced to apply the brakes which produce a retardation of 1.6 ms$^{-2}$. Find the total time taken from rest to rest.

## 4.3 Vertical motion under gravity

It is an observed fact that particles free to move in a vertical direction near the earth have the same constant acceleration, denoted by $g$ (9.8 ms$^{-2}$) downwards. Therefore, all problems involving such motion can be solved by the methods described above. If $s$ is measured upwards from the point of projection then the previous formulae hold with $a = -g$, whilst if $s$ is measured downwards $a$ has to be replaced by $g$. It does not matter whether the upwards or downwards direction is taken as positive. Normally it is more sensible to measure $s$ upwards for particles projected upwards, and downwards for those released from rest. Anything projected up will, of course, come down and in such cases, taking the positive direction of $s$ to be upwards for the complete motion avoids sign errors.

A particle projected up with speed $u$ will, by equation 3, have a speed of $v = \sqrt{u^2 - 2gs}$ when at a height $s$. Therefore, the maximum height $h$ reached will be when the speed, $v$, is zero, that is when

$$u^2 = 2gh \text{ or } h = \frac{u^2}{2g}. \qquad \ldots\ldots (5)$$

Therefore, a particle projected upwards with a speed $u$ will rise a distance $u^2/2g$. Its speed on the downward path at the point of projection i.e. $s = 0$ is $v = \sqrt{u^2}$ and is again $u$. So, the velocity when it returns to the point of projection is equal in magnitude but opposite in direction to the original velocity of projection.

### Example 4.10

A stone is thrown vertically up with speed 20 ms$^{-1}$ from the top of a building 22.5 m high. After what time and with what speed will it strike the ground?

The upward displacement $s$ m at time $t$ s is, using equation 2,

$$s = 20t - \frac{1}{2}9.8\,t^2.$$

It is required to find the time when the stone is a distance 22.5 m below its original position, that is, when $s = -22.5$. Therefore,

$$-22.5 = 20t - 4.9\,t^2, \quad \text{that is,} \quad 4.9\,t^2 - 20t - 22.5 = 0.$$

The left hand side factorises as $(t - 5)(4.9t + 4.5)$, the positive root of the equation is 5, so the time taken is 5 s. The speed $v$ when $s = -22.5$ can be found by substitution in equation 3 with $a = -9.8$, giving

$$v^2 = 400 + 441,$$

giving the speed as 29 ms$^{-1}$.

## Kinematics of Rectilinear Motion

### Example 4.11

A particle is projected vertically upwards with a speed of 14 ms$^{-1}$. Find the maximum height reached and the time taken before the particle returns to the point of projection.

The maximum height is given from equation 5 as $196/19.6 = 5$ m, and the displacement $s$ m from the point of projection at time $t$ s is $s = 14t - 4.9t^2$. It returns to the projection point when $s = 0$, that is, after $\frac{14}{4.9} = 2.86$ s. The other solution corresponding to $s = 0$ is $t = 0$, which represents the starting time.

### Exercises 4.2

Take $g = 9.8$ ms$^{-2}$ throughout this exercise. Questions 1 to 5 refer to a particle projected vertically upwards from a point $O$ with speed $u$ ms$^{-1}$.

1  $u = 20$; find the height reached and the time taken to reach the ground again.

2  The maximum height reached is 45 m. Find $u$ and the time taken to first reach a height of 40 m above $O$.

3  $u = 45$; find the times at which (a) the particle speed is 35 ms$^{-1}$ (b) the particle is at a height 90 m above $O$.

4  It is found that the two times at which the particle is 35 m above $O$ differ by 6 s. Find $u$.

5  $u = 25$; find the time taken for the particle to reach 30 m below $O$.

Questions 6 and 7 refer to a particle projected vertically downwards from $O$ with speed $u$ ms$^{-1}$.

6  $u = 5$; find the speed when the particle has dropped 10 m and the time taken to reach this position.

7  $u = 4$; find the distance fallen in 9 s.

8  A particle is projected upwards from a point $O$ with speed $u$, and at time $T$ later, a second particle is projected upwards from $O$ with the same speed. Find the time that passes after the initial projection before they meet.

9  A ball is thrown vertically upwards with a speed of 12 ms$^{-1}$ from a point at a height of 2 m above the ground. Find the speed with which it reaches the ground. Given that the ball bounces directly upwards with half the speed with which it hits the floor find the height to which it first bounces.

10  A ball thrown vertically upwards with a speed of 20 ms$^{-1}$ hits the ground 5 seconds later. Find the height above the ground from which it was thrown.

11  A ball is dropped from the top of a building of height 24 m and, at the same instant, another ball is thrown vertically upwards from the ground so as to hit the other ball. The initial speed of this second ball is 12 ms$^{-1}$. Find

(i) the time when the balls collide,

(ii) the height at which they collide.

**12** A stone is dropped from the top of a high building. The stone takes 1 s to drop from the ninth floor to the eighth and 0.5 s to fall from the eighth floor to the seventh. Find the distance between the floors, assuming that it is constant.

## 4.4 Proofs of some basic results

If the acceleration has a constant value $a$ then the definition of acceleration as a derivative gives
$$\frac{dv}{dt} = a.$$
This equation can be integrated to give
$$v = at + \text{constant}.$$
If the velocity at $t = 0$ is $u$ then substituting $t = 0$ in the above gives the constant as $u$, so that
$$v = u + at,$$
which is equation 1.

Using the definition of velocity as a derivative this equation can be rewritten as
$$\frac{dx}{dt} = u + at,$$
This equation can be integrated to give
$$x = ut + \frac{1}{2}at^2 + \text{constant}.$$
If $x$ is measured from the position when $t = 0$ and denoted by $s$ then substitution in the above equation gives the constant to be zero and
$$s = ut + \frac{1}{2}at^2,$$
which is equation 2.

Squaring equation 1 gives
$$v^2 = u^2 + 2uat + a^2t^2,$$
and substituting into this equation from equation 2 gives equation 3.

Also equation 2 can be rearranged as
$$s = \frac{1}{2}(2ut + at^2) = \frac{1}{2}t(u + u + at) = \frac{1}{2}t(u + v),$$
which is equation 4.

The above derivation is the simplest type of example that you can get of solving differential equations. One important point to notice is that every integration brings in an arbitrary constant and each constant has to be found from initial conditions.

Integrating the equation
$$v = \frac{dx}{dt},$$
between two values of $t$ ($t_1$ and $t_2$, for example) gives
$$[x]_{\text{at } t=t_2} - [x]_{\text{at } t=t_1} = \int_{t_1}^{t_2} v\, dt .$$

The left hand side is the distance between the two points corresponding to $t = t_2$ and $t = t_1$ and the right hand side is the area under the $v$ - $t$ diagram between $t = t_2$ and $t = t_1$, as shown in the diagram below.

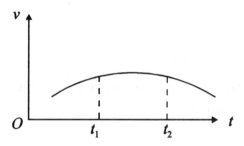

This confirms the statement made in 4.2 that the area under the $v$ - $t$ diagram corresponds to the total distance travelled.

## 4.5 Problems needing the use of calculus

The basic relations between displacement $x$, velocity $v$ and acceleration $a$ are
$$v = \frac{dx}{dt},$$
$$a = \frac{dv}{dt} = \frac{d^2x}{dt^2}.$$

The simplest types of problems that can arise are those when $x$ is given in terms of $t$ and it is required to find $v$ and/or $a$, or $v$ is given in terms of $t$ and $a$ has to be found. Solutions to problems of this type only require differentiation of the given quantities.

A slightly more complicated class of problems is that where $a$ is given in terms of $t$ and it is required to find $v$ and/or $x$, or $v$ is given in terms of $t$ and $x$ has to be found. In these cases the given quantities have to be integrated with respect to time. Every integration produces a constant of integration so the values of $x$ and/or $v$ must be known for some value of $t$.

# Kinematics of Rectilinear Motion

In dynamics, as you will find in the next chapter, force is directly proportional to acceleration. Therefore problems involving the action of forces normally mean having to find $v$ and/or $x$ from a given acceleration. In practical circumstances the acceleration may not be a function of $t$ only but may depend on $x$ and/or $v$. Problems of this type are harder to solve and will be considered in M2.

### Example 4.12
The displacement of a particle at time $t$ s is $(t^3 + 4t^2 + 6t)$ m; find its velocity and acceleration.
Differentiation with respect to $t$ gives the velocity as $(3t^2 + 8t + 6)$ ms$^{-1}$ and a further differentiation gives the acceleration as $(6t + 8)$ ms$^{-2}$.

### Example 4.13
Find the acceleration of a particle at time $t$ s given that its velocity is $t \sin 3t$ ms$^{-1}$.
The acceleration is found by differentiating the velocity with respect to time. In this case the product rule for differentiation has to be used so the acceleration is
$(\sin 3t + 3t \cos 3t)$ ms$^{-2}$.

### Example 4.14
Find the position of a particle at time $t$ s given that its acceleration is $t$ ms$^{-2}$ and that at time $t = 0$ the displacement of the particle from a fixed point and the velocity of the particle are 3 m and 6 ms$^{-1}$, respectively.
In this case
$$\frac{dv}{dt} = t,$$
where the velocity is $v$ ms$^{-1}$, and integrating this equation with respect to $t$ gives $v = \frac{1}{2}t^2$ + constant. Since the velocity is 6 ms$^{-1}$ at time $t = 0$ this gives the constant to be 6. Therefore
$$\frac{dx}{dt} = \frac{1}{2}t^2 + 6,$$
where $x$ m denotes the displacement. Integrating this equation with respect to $t$ gives $x = \frac{1}{6}t^3 + 6t$ + constant. Since the displacement is 3 m at time $t = 0$ substitution of $t = 0$ into the expression for $x$ gives the constant as 3.
Therefore the displacement is $(\frac{1}{6}t^3 + 6t + 3)$ m.

## Kinematics of Rectilinear Motion

An alternative method which would avoid introducing a constant is to integrate between $t = 0$ and $t = t$, this gives

$$[x]_{\text{at } t} - [x]_{\text{at } t = 0} = \frac{1}{6}t^3 + 6t,$$

substituting for $[x]_{\text{at } t = 0}$ as 3 gives the previous result.

The method to be used in problems where the acceleration is given in terms of $t$ essentially consists, as in the above example, of two steps:-
(i) Integrate with respect to $t$ to find $v$, remembering to introduce an arbitrary constant or to integrate between limits if $v$ is given for some value of $t$.
(ii) Integrate the expression for $v$ to find $x$, remembering to introduce a second constant or to integrate between appropriate limits if possible. Then find the constants using the given initial conditions.

### Example 4.15
A particle moving under an acceleration of $t^2$ ms$^{-2}$ at time $t$ s has a velocity of 1 ms$^{-1}$ when $t = 3$ and its displacement from a given point is 4 m when $t = 12$. Find its displacement from the given point when $t = 9$.

The velocity $v$ ms$^{-1}$ at time $t$ s satisfies the equation
$$\frac{dv}{dt} = t^2.$$
Integrating this with respect to $t$ gives $v = \frac{1}{3}t^3 + c$, where $c$ is a constant. Substituting $v = 1$ and $t = 3$ in this equation gives $1 = 9 + c$, so that $c = -8$. Therefore
$$\frac{dx}{dt} = \frac{1}{3}t^3 - 8,$$
where $x$ m denotes the displacement. This equation needs to be integrated again. One method is to integrate directly introducing another arbitrary constant and find this by using the value of $x$ at $t = 12$. The alternative, which is the one used here, is to integrate from $t = 12$ to $t = t$
This gives

$$[x]_{\text{at } t} - [x]_{\text{at } t = 12} = \frac{1}{12}(t^4 - 12^4) - 8(t - 12).$$

Substituting for the value of $x$ at $t = 12$ and evaluating $x$ at $t = 9$ gives the displacement as $-1157.25$ m.

In slightly more complicated problems, the form of the acceleration may differ from one time interval to another. In such cases, the general solutions should be found for each of the time intervals separately and the constants determined from the given conditions. The values of $x$ and $v$ at the end of one time interval may be needed to work out the appropriate constants in the succeeding interval.

### Example 4.16
A particle starts from rest with acceleration $(2 + 6t)$ ms$^{-2}$ at time $t$ s and after 2 s the acceleration changes to the constant value of 14 ms$^{-2}$ and is then maintained at this value. Find the distance covered in the first 5 s of the motion.

For the first 2 s,
$$\frac{dv}{dt} = 2 + 6t.$$
Integrating with respect to $t$ gives $v = 2t + 3t^2 + b$, where $b$ is a constant. Substituting the values at $t = 0$ gives $b = 0$ and so $v = 2t + 3t^2$. A second integration with respect to $t$ gives
$$x = t^2 + t^3 + c,$$
where $c$ is a further constant. At $t = 0$, $x = 0$, and so $c = 0$ and so $x = t^2 + t^3$. So after 2 s the particle has velocity 16 ms$^{-1}$ and has covered a distance of 12 m. It now moves with a constant acceleration of 14 ms$^{-2}$ with initial speed 16 ms$^{-1}$ for a further time of $(5 - 2)$ s $= 3$ s. Equation 2 gives
$$s = 16 \times 3 + \frac{1}{2} \times 14 \times 9 = 111$$
Therefore the required displacement is $(12 + 111)$ m $= 123$ m.

### Exercises 4.3
The following questions refer to a particle moving along $Ox$, where $x$ m denotes the displacement from $O$ of the particle at time $t$ s, and $v$ ms$^{-1}$ and $a$ ms$^{-2}$ denote the velocity and acceleration in the sense of increasing $x$, at time $t$ s.

**1** $x = 7t^4 + 2t^3 + 5$, find $a$.
**2** $v = 3t^3 + 4t^2 + 1$, find $a$.
**3** $a = 24t^2 + 18t + 2$, $v = 2$ and $x = 0$ when $t = 0$; find $x$.
**4** $a = 20t^3 + 12t^2$, $x = 1$ when $t = 0$, $v = 6$ when $t = 1$; find $x$.
**5** $a = e^{-t}$, $x = 2$ and $v = 4$ when $t = 0$; find $x$.

**6** $a = 6t$ when $0 \leq t \leq 1$, $a = 6$ when $t \geq 1$, $v = 1$ and $x = 0$ when $t = 0$; find $x$ for $t = 1$ and $t = 2$.

**7** $a = 24t^2 + 6$ when $0 \leq t \leq 1$, $a = 30t$, $t \geq 1$, $v = 2$ and $x = 1$ when $t = 0$; find $x$ for $t > 1$.

## 4.6 Modelling considerations

The assumption of starting from rest with a constant acceleration is not particularly valid as in practical circumstances the acceleration would build up very quickly from zero to a constant value. Therefore a more refined model would take into account the starting up phase. For example, instead of assuming a particle has a constant acceleration of 1 ms$^{-2}$ for 10 s from rest, a more realistic model would be to assume a variable acceleration for the first second and then an acceleration of 1 ms$^{-2}$ for the remaining nine seconds. The basic properties of this variable acceleration are that it is zero at the start and equal to 1 ms$^{-2}$ after 1 s. You might be given a choice of various functions and asked to pick the most appropriate one or given an acceleration of the form $a + bt$ and asked to determine the constants $a$ and $b$.

In the particular example above a possible form is $0.157(e^{2t} - 1)$, where $t$ s denotes the time from the start. This is zero for $t = 0$ and builds up quickly to 1 (actually it is 1.003) when $t = 1$. The numerical constant is the approximate value of $1/(e^2 - 1)$. If $v$ ms$^{-1}$ denotes the velocity at time $t$ s then

$$\frac{dv}{dt} = 0.157(e^{2t} - 1).$$

Integrating this equation from $t = 0$ to $t = 1$ gives

$$[v]_{\text{at } t=1} - [v]_{\text{at } t=0} = \int_0^1 0.157(e^{2t} - 1)\, dt \ .$$

The numerical value of the right hand side is 0.34, the velocity at $t = 0$ is zero and therefore the velocity when $t = 1$ is 0.34 ms$^{-1}$. The acceleration for the next nine seconds is 1 ms$^{-2}$ and therefore, from equation 1, the velocity at the end of this period is $(0.34 + 1 \times 9)$ ms$^{-1}$. If the acceleration had been taken as 1 ms$^{-2}$ throughout then the velocity after 10 s would have been 10 ms$^{-1}$.

It is often necessary in modelling situations to take a simple form, such as $a + bt$, for a velocity or acceleration and then find the constants so as to satisfy particular conditions.

## Kinematics of Rectilinear Motion

**Miscellaneous Exercises 4**

**1**

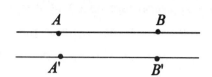

The diagram shows two parallel tracks for toy cars, the segments $AB$ and $A'B'$ are each of length 8.2 m. At time $t = 0$ a toy car $P$ passes through $A$ with speed 0.8 ms$^{-1}$ and moves towards $B$ with constant acceleration of 0.3 ms$^{-2}$ in the sense from $A$ to $B$. At time $t = 0$ a toy car $Q$ passes through $B'$ with speed 0.4 ms$^{-1}$ and moves towards $A'$ with constant acceleration of 0.1 ms$^{-2}$ in the sense from $B'$ to $A'$. Modelling the cars as particles find
(a) the speed of $P$ when $t = 1$,
(b) the distance of $P$ from $A$ when $t = 1$,
(c) an expression for $PQ$ at time $t$ s,
(d) the times when $P$ and $Q$ are at a distance of 2.8 m apart.
(In parts (c) and (d) the perpendicular distance between the tracks is to be neglected)

**2** A particle $P$ is projected along rough horizontal ground from a fixed point $A$ towards a second fixed point $B$, where $AB = 60$ m, with speed 4 ms$^{-1}$ and moves under a constant frictional retardation of $\frac{1}{3}$ ms$^{-2}$. At the same instant a particle $Q$ is projected along the ground from $B$ towards $A$ with speed $v$ ms$^{-1}$ and is subject to a constant frictional retardation of 2 ms$^{-2}$. Show that $P$ will come to rest after 12 s at a distance of 24 m from $A$. Hence deduce that the particles collide if $v \geq 12$.
Show that for the case $v = 13$, the collision occurs after 6 s.
Find the value of $v$ such that the collision occurs after 4 s and determine the distance from $A$ at which the collision takes place.

**3** A car passes a point $A$ with speed 35 ms$^{-1}$ and continues at the same speed. Two seconds later a police motor-cyclist sets off from $A$. He accelerates at a rate of 6 ms$^{-2}$ until he reaches a speed of 45 ms$^{-1}$ and then moves at this speed until level with the car. Find the distance from $A$ to the point where the motor-cycle and car are level.

**4** A cheetah is estimated to be able to run at a maximum speed of 100 kmh$^{-1}$ whilst an antelope can run at a maximum speed of 65 kmh$^{-1}$. A cheetah at rest sees an antelope at rest, and starts running towards it. The antelope immediately starts moving away. Both animals are assumed to move with constant acceleration and reach their maximum speeds in 4 s. Assuming that both run along the same straight line and that the cheetah catches up with the antelope in 15 s find the distance between the animals when they first started moving.

**5** A cage goes down a mine shaft 480 m deep in 45 s. The cage is accelerated uniformly from rest for the first 160 m, it then travels at a constant speed of $v$ ms$^{-1}$ for the next 240 m; finally it is brought to rest with a uniform retardation over the last part of its journey. Find

(i) $v$,

(ii) the acceleration of the cage on the first part of its journey.

**6** A train starting from rest moves with constant acceleration during the first 50 seconds of its journey when it covers 500 m. It then travels at constant speed until it is brought to rest by applying a constant retardation. The distance travelled whilst the retardation is applied is 800 m. By drawing a speed-time graph, or otherwise, find

(i) the maximum speed attained by the train,

(ii) the magnitude of the retardation.

Given that the total journey time was 4 minutes, determine the distance covered at constant speed.

**7** A particle moves with constant acceleration from $A$ to $B$, where $AB = 96$ m, and then moves with constant retardation from $B$ to $C$, where $BC = 30$ m. The speeds of the particle at $A$ and $B$ are 6 ms$^{-1}$ and $u$ ms$^{-1}$ respectively and it comes to rest at $C$. Find, in terms of $u$, the times taken by the particle to move from $A$ to $B$ and from $B$ to $C$.

Given that the total time taken by the particle to move from $A$ to $C$ is 18 seconds, find

(i) the value of $u$,

(ii) the acceleration and the retardation of the particle.

**8** A train driver, in order to satisfy speed restrictions on a length of track, has to apply a constant retardation of 1 ms$^{-2}$ in order to reduce the speed from 40 ms$^{-1}$ at a point $A$ to a speed of 10 ms$^{-1}$ at another point $B$. The train travels from $B$ to $C$, a distance of 3.5 km, at a constant speed of 10 ms$^{-1}$ and then moves with constant acceleration of 0.2 ms$^{-2}$ so that its speed at a point $D$ is 40 ms$^{-1}$. Sketch the velocity-time graph for the journey from $A$ to $D$, and show that the distance from $A$ to $D$ is 8 km.

Show that the journey from $A$ to $D$ takes 330 s more than it would if the train travelled at a constant speed of 40 ms$^{-1}$ from $A$ to $D$.

**9** A lift travels vertically a distance of 22 m from rest at the basement to rest at the top floor. Initially the lift moves with constant acceleration for a distance of 5m; it then continues with constant speed $u$ ms$^{-1}$ for 14m, a constant retardation is then applied so that the lift comes to rest at the top floor. Find, in terms of $u$, the times taken by the lift to cover the three stages of its journey.

Given that the total time that the lift is moving is 6 seconds, find
(i) the value of $u$,
(ii) the values of the acceleration and retardation.

**10** A car starting from rest at a point $A$ moves along a straight road with constant acceleration $f$ until it reaches a speed $v$; it then continues at this speed. When the car starts, a second car is at a distance $b$ behind the first car and moving in the same direction with constant speed $u$. Find the distance between the cars at time $t$ after the first car has left $A$ for
(i) for $0 < t < v/f$,
(ii) for $t > v/f$.
Show that the second car cannot overtake the first one during the period $0 < t < v/f$ unless $u^2 > 2fb$.
Find the least distance between the two cars in the case $u^2 < 2fb$ and $u < v$. State briefly what will happen if $u^2 < 2fb$ and $u > v$.

**11** A tube train travels between the two stations at a distance of 2.7 km apart. The train either moves with constant acceleration of 0.25 ms$^{-2}$ or stays at a constant speed or moves with constant retardation of 0.5 ms$^{-2}$. The train is subject to a maximum speed limit of 15 ms$^{-1}$. Given that the train completes the journey, from rest to rest, in the shortest possible time without exceeding the speed limit, find the time taken and show that the distance covered at maximum speed is 2.025 km.

**12** A railway timetabler has to determine the time taken between two stations a distance of 22.5 km apart. In order to do this he assumes that a train, on leaving one station, accelerates at a constant rate for 75 s until it reaches a constant speed of 30 ms$^{-1}$ at a point $A$. It then continues at this speed to point $B$ when a constant retardation is applied for 75 s so that the train comes to rest in the second station.
(a) Draw a sketch of the velocity-time graph showing the motion of the train.
(b) Find the magnitude of the acceleration and the total distance travelled whilst the train is accelerating,
(c) Find the total time estimated by the timetabler for the journey.
(d) The timetabler realises that in making a timetable he has to insert a safety margin by allowing for the train having to stop at a signal. He does this by assuming that some time after passing $A$ the train is forced to start slowing down to rest and then to stop for 60 s. He assumes that the retardation and acceleration applied are the same as those applied at the start and end of the journey. He also assumes that the point where the train starts slowing down is such that the train will have resumed the constant speed of 30 ms$^{-1}$ before reaching $B$. Find the total time that would now be estimated for the journey.

**13** A motorist travelling at 100 kmh$^{-1}$ sees an obstruction 100 m ahead. After a delay of 0.3 s he applies the brakes and immediately starts to decelerate at a rate of 4 ms$^{-2}$. How far from the obstruction does he stop?

**14** When a car driver sees an obstruction ahead, there is a delay of $T$ s (the driver's reaction-time) before he takes any action. The brakes are then applied so that the car moves with constant retardation $f$ ms$^{-2}$. When the driver sees an obstruction when travelling at a uniform speed of 12 ms$^{-1}$, he can bring the car to rest in 20 m. If he is travelling at a uniform speed of 24 ms$^{-1}$ when he sees the obstruction he can bring the car to rest in 64 m. Find $f$ and $T$.

The driver sees an obstruction when travelling at 18 ms$^{-1}$ with constant acceleration of 3 ms$^{-1}$. Show that he can stop the car in 46 m.

**15** A ball is projected vertically upwards with an initial speed of 14.7 ms$^{-1}$. Find

(i) its greatest height,

(ii) the time taken to reach its greatest height,

(iii) its speed two seconds after projection.

**16**

The diagram shows a girl rolling a ball up a slope. The slope is such that the ball experiences a retardation of 4 ms$^{-2}$. Given that the ball starts moving up the slope with speed 2 ms$^{-1}$ find

(i) the distance it moves up the plane before it comes to instantaneous rest,

(ii) the total time before the ball returns to the girl.

**17** At time $t = 0$ a particle $A$ is projected vertically upwards from the point $O$ with speed $U$. At time $t = U/2g$ a second particle $B$ is projected vertically upwards from $O$ with speed $3U$. Show that the particles collide before $A$ has reached its maximum height.

Find the height above $O$ at which the particles collide.

The particle $A$ is then projected for a second time from $O$ and a third particle $C$ is projected vertically upwards from $O$ at a time $U/2g$ later. This collides with $A$ at time $U/g$, after $A$ was projected. Find the speed of projection of $C$.

**18** At time $t = 0$ a particle is projected vertically upwards from $O$ with speed 19.6 ms$^{-1}$ and, two seconds later, a second particle is projected from $O$ with the same speed. Express the heights of both particles above $O$ in terms of $t$ and hence, or otherwise, find the value of $t$ when they collide.

Find the speeds of the particles at the instant of collision.

**19** Two free-falling raindrops leave the top of a cliff of height $h$ such that the second one begins to fall when the first one has already fallen a distance $s$. Show that the distance between the drops when the first drop hits the ground is $2\sqrt{sh} - s$.
If the height of the cliff is 28 m and this final distance apart is 3 m, find the value of $s$ to the nearest centimetre. Take $g$ as 9.8 ms$^{-2}$.

**20** A particle $P$ moves along the $x$-axis so that its velocity at time $t$ s is $v$ ms$^{-1}$ where
$$v = 9t^2 - 4t + 1.$$
Given that $P$ is at the origin when $t = 0$, find
(a) the distance of $P$ from the origin when $t = 1$,
(b) the acceleration of $P$ when $t = 1$.

**21** A particle moves along a straight line so that its acceleration at time $t$ seconds is $(6t - 8)$ ms$^{-2}$. At $t = 0$ seconds the particle passes through the fixed point $O$ with a velocity 4 ms$^{-1}$. Find
(i) the distance from $O$ of the point where the particle first comes to instantaneous rest,
(ii) the total time $T$ seconds taken by the particle to return to the starting point,
(iii) the greatest speed of the particle for $0 < t < T$.

**22** A particle moves in a straight line so that its speed at time $t$ s is inversely proportional to $(t + 3)$, and when $t = 2$ s, the particle has a retardation of $4/25$ ms$^{-2}$. Given that the particle is at $O$ at time $t = 0$ find its distance from $O$ when $t = 1$.

**23** The acceleration at time $t$, of a particle moving in a straight line is $k \sin pt$, where $k$ and $p$ are constants. At time $t = 0$ the particle is at the point $O$ and moving with velocity $u$. Show that its velocity at any subsequent time is
$$u + \frac{k}{p}(1 - \cos pt).$$
Show that, for $u = 0$, the particle first comes to instantaneous rest after travelling a distance $\frac{2\pi k}{p^2}$.

**24** A particle, moving in a straight line, starts from rest at time $t = 0$ s and at time $t$ s its velocity $v$ ms$^{-1}$ is given by
$$v = 3t(t - 4) \text{ for } 0 \le t \le 5, \quad v = 75/t \quad 5 \le t \le k,$$
where $k$ is a constant.
(i) Sketch the velocity-time diagram for the particle for $0 \le t \le k$.
(ii) Find the range of values of $t$ for which the acceleration of the particle is positive.
(iii) Show that the **total** distance covered by the particle in the interval $0 \le t \le 5$ is 39 metres.

(iv) Given that the distance covered by the particle in the interval $5 \le t \le k$ is also 39 metres, find, to 2 significant figures, the value of $k$.

**25** In order to model the final stages of the motion of a bird it is assumed that its speed is $(a + bt)$ ms$^{-1}$, where $a$ and $b$ are constants. The speed of the bird when $t = 0$ is 4 ms$^{-1}$ and it comes to rest when $t = 3$. Find the values of $a$ and $b$.

A more refined model is then sought which is such that the acceleration of the bird is zero when it comes to rest. Assuming that, in this case, $v = p + qt + rt^2$, where $p$, $q$ and $r$ are constants find the values of these constants.

Determine, for this model, the distance travelled by the bird in the last three seconds of its flight.

# Chapter 5

# Dynamics of Rectilinear Motion

After working through this chapter you should
- know Newton's laws of motion and be able to apply them to determine the rectilinear motion of bodies under the action of given forces,
- be able to solve simple problems involving the motion of connected particles,
- be able to solve simple problems involving the motion of vehicles.

## 5.1 Newton's laws of motion

**First law**

This law states that every body continues in a state of rest or of uniform motion in a straight line unless compelled to change that state by a force. If a body changes from a state of uniform motion, that body has an acceleration (or retardation) and, therefore, Newton's first law can be interpreted as stating that a force acting on a body produces an acceleration (or retardation) and that any body which possesses an acceleration (or retardation) is being acted upon by a force.

**Second law**

This second law needs the introduction of a new fundamental quantity, the mass. You will probably have used the idea of mass of a body though you may not have a clear idea of what it means. The theoretical method of defining and measuring mass is given in 5.6.
Mass is often defined as the quantity of matter in a body but this is not a particularly good definition. The important point is that the mass of a body is a fundamental property of that body and that the unit of mass can be defined independently of units of time, length or force. The unit used in the S.I. system is the kilogram (kg), and 1000 kg is one tonne. Newton's second law states that the component of a force in a given direction acting on a body is equal to the product of the mass of the body and the acceleration in the given

direction. The only problems considered will be those when the acceleration is along a fixed direction. Motion will not be possible in any other direction, therefore there can be no acceleration in any other direction and therefore the forces in any other direction are in equilibrium. Symbolically

$$F = kma,$$

where $a$ is the acceleration in a reference direction along the line, $F$ is the component of the force along the line in the reference direction and $k$ is a positive constant. It is very important always to remember to choose a particular reference direction at the start of any problem.

In the S.I. unit system the unit of force (the newton) is chosen so that $k = 1$, so that one newton produces unit acceleration (1 ms$^{-2}$) when acting on unit mass (1 kg). In this system the above equation becomes

$$F = ma.$$

You will come across problems where not all the forces are in the fixed direction of motion. In this case, as mentioned above, the forces in any other direction are in equilibrium and the equation of motion has to be supplemented by this additional condition.

**Third law**

Newton's third law states that action and reaction are equal and opposite. You have already met this in Statics but it also applies to bodies in motion. In the diagram if the car exerts a force to the right and of magnitude $F$ on the caravan then the caravan exerts a force of magnitude $F$ to the left on the car.

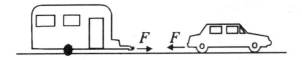

## 5.2 Motion of single particles under the action of constant forces

It is very important in order to set up a correct equation of motion to make a clear sketch showing the position of the particle at any time and the reference direction and to mark, as in problems in Statics, all the forces acting. The next stage, in most problems, is to find the component of the force in the reference direction and then Newton's second law can be used to find the acceleration. For constant forces, the position of a particle at any time can be found by using the formulae in chapter 4.

# Dynamics of Rectilinear Motion

In some problems the direction of motion may not be obvious. This would occur for example when there are two or more tugs pulling a ship from different directions. In this case the resultant would have to be found and the magnitude and direction of the acceleration then determined by Newton's second law.

Ocassionally the forces are not known but the motion of a particle can be observed so that the acceleration can be calculated and hence the force determined.

### Example 5.1

A stone of mass 0.5 kg is falling through water. The buoyancy of the water provides an upward force of 3 N and the resistance provides a further upward force of 1 N. Find the acceleration of the stone.

The reference direction is taken to be vertically downwards. The vertical forces acting are the force of gravity of 0.5g downwards, the resistance and the buoyancy. These forces are shown in the diagram.

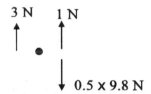

The total force in the downward direction is $(0.5 \times 9.8 - 3 - 1)$ N $= 0.9$ N and if the downward acceleration is $a$ ms$^{-2}$ then Newton's second law gives
$$0.5\,a = 0.9 \text{ N},$$
and so the acceleration is 1.8 ms$^{-2}$.

### Example 5.2

A toboggan of mass 60 kg is being pulled on horizontal snow by a boy exerting a horizontal force of 100 N and it moves with acceleration 1.5 ms$^{-2}$. Find the frictional resistance to the toboggan.

The reference direction is chosen to be that in which the boy is pulling. The only other force acting on the toboggan is the frictional resistance $F$ N, which will be in the opposite direction to that in which the boy is pulling. The forces are shown in the diagram below.

# Dynamics of Rectilinear Motion

The total force to the right is, therefore, $(100 - F)$ N. Therefore Newton's law gives
$$100 - F = 60 \times 1.5,$$
so that the resistance is 10 N.

## Example 5.3
A man of mass 70 kg is in a lift. Find the forces exerted by the floor of the lift on the man
(i) when the lift descends with an acceleration of 0.3 ms$^{-2}$,
(ii) when on its downward journey, the lift is slowing with a retardation of 0.3 ms$^{-2}$,
(iii) when the lift is descending with acceleration $g$.

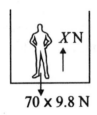

The reference direction is taken vertcally downwards, and the forces are as shown in the diagram.

(i) The only forces acting on the man are the force of gravity down and the unknown reaction of the floor, of magnitude $X$ N, acting up. The total downward force in newtons is, therefore, $(70 \times 9.8 - X)$ N. Therefore
$$70 \times 0.3 = 70 \times 9.8 - X,$$
Hence the force is 665 N.

So the force exerted by the floor on the man is 665 N which is less than the man's weight of $70 \times 9.8$ (= 686) N. This would be demonstrated in practice by a slight lightening effect on the legs of the man when the lift starts to accelerate downwards. If the man had been standing on bathroom scales in the lift then the force would be the reaction of the scales on the man and this, as discussed in 2.4, would be the reading on the scales and therefore his weight would have apparently been changed.

(ii) A retardation of 0.3 ms$^{-2}$ means that the downward acceleration is $-$ 0.3 ms$^{-2}$ and, therefore, still taking the downward direction to be the reference direction, with the upward force exerted on the man now being $Y$ N, Newton's second law gives
$$-70 \times 0.3 = 70 \times 9.8 - Y,$$
the reaction is now 707 N and there is an extra thrust on the legs as the lift slows down when going downwards. This means that the man's apparent weight would have increased.

(iii) Replacing 0.3 by 9.8 in the equation of motion in (i) gives
$$70 \times 9.8 = 70 \times 9.8 - X,$$
This gives $X = 0$ so that the man would appear weightless! Effectively the lift would be in 'free fall' and this would only occur if the cable broke.

## Example 5.4

A particle of mass 0.3 kg is projected up a line of greatest slope of a smooth plane inclined at an angle of $30°$ to the horizontal. Given that its initial speed is 19.6 ms$^{-1}$ find how far up the plane it travels before coming to instantaneous rest.

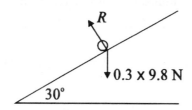

The forces acting on the particle are shown in the diagram and this is a simple example of the situation when there are forces acting in directions other than that of the motion. The force due to gravity has a component $0.3 \times 9.8 \times \sin 30°$ N acting down the plane and one $0.3 \times 9.8 \times \cos 30°$ N perpendicular to the plane. The forces off the line of greatest slope are in equilibrium and therefore the reaction $R$ of the plane is equal to $0.3 \times 9.8 \times \cos 30°$ N. Taking the reference direction for acceleration up the plane Newton's law shows that the acceleration, $a$ ms$^{-2}$, satisfies
$$0.3 a = -0.3 \times 9.8 \times \sin 30°,$$
so that the acceleration is $-4.9$ ms$^{-2}$, i.e. it is a retardation.
The distance up the plane can now be found from equation 3 of Chapter 4 ($v^2 = u^2 + 2as$) so that
$$0 = 19.6^2 - 9.8s,$$
showing that the distance travelled up the plane before the particle comes to instantaneous rest is 39.2 m.

## Example 5.5

A particle of mass 0.5 kg is sliding down a rough plane inclined at an angle $\theta$ to the horizontal where $\cos \theta = \frac{3}{5}$ and $\sin \theta = \frac{4}{5}$. The coefficient of (sliding) friction between the plane and the particle is 0.5. Find the acceleration of the particle.

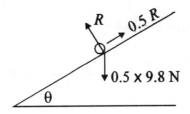

The forces acting are shown in the diagram. Since the particle is sliding down the plane the friction force will be up the plane and of magnitude $0.5\,R$, where $R$ denotes the normal reaction of the plane. The force of gravity has a component down the plane of magnitude $0.5 \times \sin\theta \times 9.8$ N and $0.5 \times \cos\theta \times 9.8$ N perpendicular to the plane. As in the previous example the forces not along the plane are in equilibrium so the reaction of the plane is $0.5 \times \cos\theta \times 9.8$ N $= 2.94$ N. This gives the friction force as $1.47$ N. The acceleration $a$ ms$^{-2}$ down the plane therefore satisfies

$$0.5\,a = 0.5 \times \sin\theta \times 9.8 - 1.47,$$

giving the acceleration down the plane as $4.9$ ms$^{-2}$.

### Example 5.6

The diagram shows two tugs towing a tanker of mass 20 000 tonnes, the direction and magnitude of the forces exerted by the tugs are shown. Find the magnitude and direction of the acceleration of the tanker.

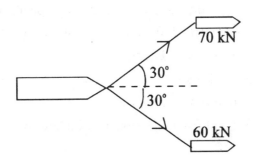

The first step is determining the resultant of the forces. This has a component of
$(60 \cos 30° + 70 \cos 30°)$ kN $= 112.58$ kN  along the dotted line and
$(70 \sin 30° - 60 \sin 30°)$ kN $= 5$ kN  perpendicular to the dotted line and up the page. The resultant is therefore of magnitude $\sqrt{112.58^2 + 5^2}$ kN $= 112.7$ kN  at an angle $\theta$ above the dotted line where $\tan\theta = \dfrac{5}{112.58} = .044$.

Therefore $\theta$ is $2.5°$ and the acceleration is of magnitude $\dfrac{112.7}{20}$ ms$^{-2}$ $= 5.64$ ms$^{-2}$ at an angle of $2.5°$ above the dotted line.

## Dynamics of Rectilinear Motion

### Exercises 5.1

**1** A tug tows a barge of mass 9000 kg with acceleration 0.1 ms$^{-2}$. The water resistance to the movement of the barge is 900 N. Given that the tow rope is horizontal, calculate the tension in this rope.

**2** The resistance to the motion of a car of mass 700 kg is 400 N and the driving force exerted by the engine is 1800 N. Find the acceleration of the car.

**3** A constant retarding force of 2000 N is applied to a car of mass 800 kg moving on a level road so as to reduce its speed from 20 ms$^{-1}$ to 10 ms$^{-1}$. Find, for the interval during which the car reduces speed,

(i) the time taken,

(ii) the distance covered.

**4** A body of mass 10 kg is moved vertically upwards from rest by a constant force to a height of 5 m above its starting point and given a speed of 6 ms$^{-1}$. Find the magnitude of the force.

**5** The resistance to the motion of a car moving along a horizontal road is 1500 N. The engine of the car exerts a constant force of 5400 N and the car reaches a speed of 24 ms$^{-1}$ from rest in 8 s. Find

(i) the acceleration of the car during the first 8 s,

(ii) the mass of the car,

(iii) the distance travelled by the car during the first 8 s.

**6** A child of mass 28 kg stands in a lift moving with an upwards acceleration of 1.4 ms$^{-2}$. Find

(i) the force exerted on the child by the floor of the lift,

(ii) the force exerted on the floor by the child.

**7** A lift accelerates upwards from rest at a constant rate of 0.5 ms$^{-2}$ to a maximum speed of 2 ms$^{-1}$ and after moving at this constant speed it moves with constant retardation of 0.25 ms$^{-2}$ until it comes to rest.

A parcel of mass 10 kg is standing on the floor of the lift during this ascent from rest to rest. Find the magnitude of the force exerted by the floor of the lift on the parcel during each stage of the ascent.

**8** A man of mass 180 kg stands on bathroom scales on the floor of a lift. The lift is moving with a downwards acceleration of 2 ms$^{-2}$. What is the reading on the scales?

The man then pushes on the ceiling of the lift with a walking stick. Which of the following readings is likely to be the correct one 1420 N or 1390 N?

The man then applies the same force to the walking stick with the latter pushing the floor of the lift. Find the reading of the scales in this case.

**9** A car is moving along a horizontal road at a steady speed of 30 ms$^{-1}$. Find the ratio of the magnitude of the constant force necessary to bring the car to rest in a distance of 40 m to the magnitude of the constant force necessary to bring the car to rest in a time of 8 s.

**10** A stone is sliding over the ice on a pond and it slides a distance of 200m before coming to rest from a speed of 15 ms$^{-1}$. Find the coefficient of friction.

**11** A ring of mass 0.01 kg is threaded on a smooth fixed vertical wire. A force of 0.4 N acts on the ring and the line of action of the force makes an angle of 60° with the upward vertical. Find the magnitude of the upward acceleration of the ring.

**12** A particle of mass $m$ moves up a line of greatest slope of a plane inclined at an angle $\theta$ to the horizontal, where $\cos\theta = \frac{4}{5}$ and $\sin\theta = \frac{3}{5}$, under the action of a force $F$ of magnitude $2mg$ acting parallel to the plane. Given that the coefficient of friction between the particle and the plane is 0.5, find the acceleration of the particle.

**13** A particle of mass $m$ is projected up a line of greatest slope of a plane inclined at an angle $\theta$ to the horizontal, where $\cos\theta = \frac{4}{5}$ and $\sin\theta = \frac{3}{5}$, with speed 14.7 ms$^{-1}$.

(i) Find for the case when the plane is smooth, how far up the line of greatest slope the particle will travel.

(ii) Find, for the case when the plane is rough and the particle travels a distance of 14 m up the line of greatest slope before coming to rest, the coefficient of friction between the particle and the plane.

**14** A particle of mass 0.2 kg is pushed up a line of greatest slope of a plane which is inclined at an angle angle $\theta$ to the horizontal, where $\cos\theta = \frac{4}{5}$ and $\sin\theta = \frac{3}{5}$, by a horizontal force of magnitude 2 N. Given that the plane is smooth, find the magnitude of the acceleration of the particle.

**15**

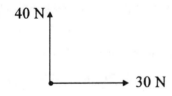

The diagram shows a particle of mass 0.5 kg on a smooth floor being pulled in two perpendicular directions by forces of magnitude 30 N and 40 N respectively. Find the direction in which the particle moves and the magnitude of its acceleration.

## 5.3 Motion of connected particles under the action of constant forces

**Motion along a single line**

The simplest type of problems with two, or more, particles connected together are those where all the particles are moving along the same line e.g. a car pulling a trailer, and therefore all particles have the same acceleration. In more complicated cases the particles are connected by a string passing over a pulley so that they are not all moving in the same straight line.

The general approach is still the same. Draw a diagram showing the reference direction and the forces acting on each particle. In these case it is necessary to include the forces between the particles and to remember that Newton's third law has to be satisfied. It is often necessary, in order to solve a particular problem, to apply Newton's second law to both particles or, alternatively, to one particle and to the system as a whole. This avoids calculating the force between the particles. If you consider just one of the particles then you have to include the force between the particles.

### Example 5.7

A car of mass 900 kg tows a caravan of mass 600 kg along a horizontal road. The driving force due to the engine is 600 N. Given that there are no resistances acting find
(i) the acceleration of the car and caravan,
(ii) the tension in the tow bar.

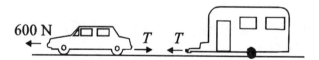

The forces acting are shown in the diagram. If the system of car and caravan is taken together the tension forces in the tow bar cancel and the total force acting is 600 N. Newton's second law gives, denoting the acceleration by $a$ ms$^{-2}$,

$$1500\,a = 600 \text{ N},$$

so that the acceleration is 0.4 ms$^{-2}$.

Newton's second law can now be applied to the caravan. The only force acting is the tension $T$, therefore

$$T = 600 \times 0.4 \text{ N} = 240 \text{ N}.$$

It would have been possible to use the equation of motion of the car. The force acting is $(600 - T)$ N and this is equal to $900 \times 0.4$ N, giving the same answer.

## Example 5.8

A car of mass 800 kg is towing a trailer of mass 100 kg up a hill inclined at an angle $\alpha$ to the horizontal where $\sin \alpha = \frac{1}{14}$. The driving force due to the engine is 900 N and the resistances to the car and trailer are, respectively, 150 N and 30 N. Find the acceleration of the car and the tension in the tow bar.

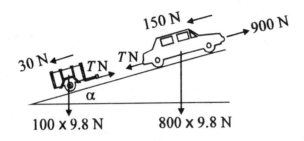

The forces acting are shown in the diagram. In this case the component of the weight down the hill has to be taken into account. Taking the car and trailer together the total resistance acting down the hill is 180 N. The component of the total weight down the hill is $900 \times 9.8 \times \sin \alpha$ N = 630 N. The total component of force, acting up the hill, is therefore $(900 - 630 - 180)$ N = 90 N. The acceleration, $a$ ms$^{-2}$, is given by

$$900a = 90,$$

so that $\qquad a = 0.1$.

Considering next the trailer. The component of the weight down the hill is

$$100 \times 9.8 \times \sin \alpha \text{ N} = 70 \text{ N},$$

the resistance is 30 N and therefore the tension, $T$ N, satisfies

$$T - 30 - 70 = 100 \times 0.1.$$

so that the tension is 110 N.

## Example 5.9

An engine of mass 50 tonnes pushes a carriage of mass 10 tonnes with an acceleration of 0.4 ms$^{-2}$. The resistance to the motion of the engine is 1200 N and to the motion of the carriage is 800 N. Find
(i) the total driving force of the engine,
(ii) the force exerted on the carriage by the engine.

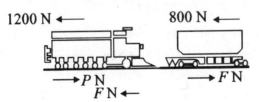

# Dynamics of Rectilinear Motion

The total force on the two is $(P - 2000)$ N to the right, where $P$ N denotes the driving force, and therefore, by Newton's second law,
$$P - 2000 = 60000 \times 0.4,$$
so that the driving force is 26 000 N.

The force acting to the right on the carriage is $(F - 800)$ N, where $F$ N is the force exerted by the engine. Therefore
$$F - 800 = 10000 \times 0.4,$$
so that the force exerted by the engine is 4800 N

## Exercises 5.2

**1** A car of mass 700 kg tows a caravan of mass 500 kg along a horizontal road. Given that the driving force is 300 N and neglecting resistances, find the acceleration of the car.

**2** A car of mass 900 kg is pulling a caravan of mass 600 kg, by means of a tow bar, along a straight horizontal road. The resistive forces opposing the motions of the car and the caravan are 120 N and 60 N respectively. Given that the car is accelerating at 2 ms$^{-2}$, find
(i) the tension in the tow bar,
(ii) the force being produced by the engine of the car.

**3** A car of mass 800 kg tows a caravan of mass 400 kg against a total resistance of 600 N. Given that the acceleration of the car is 0.8 ms$^{-2}$ and that the resistances on the car and caravan are proportional to their masses find
(i) the driving force,
(ii) the tension in the tow bar.

**4** A car of mass 800 kg has a driving force of 2.1 kN when pulling a caravan of mass 400 kg and has a constant acceleration of 0.5 ms$^{-2}$. Given that the resistances on the car and caravan are proportional to their masses find these resistances and the tension in the tow bar.

**5** A tug, of mass 50 000 kg tows three barges in line behind it along a canal. Each barge is of mass 10 000 kg and the resistance to the motion of each barge is 800 N. The resistance to the motion of the tug is 3000 N. Calculate the tensions in the tow ropes between
(i) the tug and the first barge,
(ii) the first and second barges,
(iii) the second and third barges, when
(a) the tug and barges are moving at a uniform speed,
(b) they are all accelerating at 0.5 ms$^{-2}$.

**6** A car of mass 1200 kg is pulling a caravan of mass 300 kg up a slope inclined at an angle $\alpha$ to the horizontal where $\sin \alpha = \dfrac{1}{196}$. The driving force is 1800 N and the resistances to the car and caravan are 275 N and 100 N respectively. Find the acceleration of the system and the tension in the tow bar.

## Motion involving pulleys

In some problems involving connected particles each particle is moving on a straight line but they are not moving in the same straight line. This happens when the connection (e.g a string) passes over a pulley. Some typical configurations are shown in the diagram.

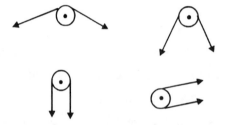

In this case the acceleration of the particles (assuming the string is inextensible) has the same magnitude but not the same direction. The same approach has again to be used i.e. showing all the forces on a clear diagram and applying Newton's law separately to each particle.

## Modelling a pulley

A pulley is effectively a device for changing the direction of a force and is normally free to turn about an axis. If the rim of the pulley is smooth then a string passed over it simply slides along it without making the pulley turn and the pulley is then effectively the same as a smooth peg. This point was discussed in the Glossary in 2.2. In practice, however, the rim of the pulley is rough so that normally a string does not slide relative to it and the string makes the pulley rotate.

It is possible to show that the difference in the tensions $T_1$ and $T_2$ at the points shown in the diagram is proportional to the mass and radius of the pulley and if these are neglected (the pulley is said to be small) then the tensions are equal. This is the normal assumption. Since the pulley can rotate then there may be friction at the axis and this also can make $T_1$ different from $T_2$. If this friction can be neglected, the pulley is described as smooth, which is slightly different from the use in Statics. To summarise, in Dynamics a string is assumed not to slide relative to the pulley and a small smooth pulley is one where the tensions $T_1$ and $T_2$, shown above, are equal and there is no frictional effect at the axle of the pulley.

**Example 5.10**

A light inextensible string passes over a small smooth pulley and particles of masses $3m$ and $m$, are attached to the ends of the string and can move, with the parts of the string not in contact with the pulley being taut and vertical. Find the acceleration of the particles and the tension in the string.

Since the string is light the tension will be constant throughout any straight length, and the fact that the pulley is small and smooth ensures the tension is the same throughout the string.

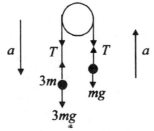

The forces acting on the particles are shown in the diagram; since the string is taut, the particles will both have accelerations of the same magnitude $a$ but in opposite directions, as shown in the diagram. The downward component of the force on the heavier particle is $3mg - T$ so that applying Newton's second law to it gives

$$3mg - T = 3ma.$$

The acceleration of the lighter particle is upwards and the upwards component of the force on it is $T - mg$. Newton's law gives

$$T - mg = ma.$$

Adding the equations gives

$$2mg = 4ma,$$

so that $a = \dfrac{g}{2}$. Substituting this into either of the other equations gives $T = \dfrac{3mg}{2}$.

It is worth noticing that the equation for *a* is precisely the equation that would have been obtained by considering the horizontal motion of the system formed by rotating clockwise one of the vertical parts of the string and the other part counterclockwise, so that the string becomes straight and horizontal with there being no change, relative to the string, in the forces acting.  In this changed problem the accelerations of both particles are in the same direction but the forces are in opposite directions.  It is certainly not a correct method to use for this problem and others involving strings over pulleys and **it should not be used in examinations** since it needs to be proved correct in each case.  It is also very easy to make a mistake and therefore marks, even when awardable, might be lost.  Also most examination questions tend to ask for the tension and this needs the writing down of at least one equation of motion.  There is therefore very little benefit in attempting to use this particular 'shortcut'.  At best it should only be used as a check on the algebra involved in eliminating the tension.

### Example 5.11

The left hand diagram below shows a particle $A$, which is of mass $10m$ and lies on a smooth rectangular table.  Particle $A$ is connected by light inextensible strings to two particles $B$ and $C$ of masses $4m$ and $2m$ respectively and passing over smooth pulleys at the opposite edges of the table.  The strings are both perpendicular to the edges of the table.  Find the acceleration of $A$, $B$ and $C$ and the tension in the string attached to $C$.

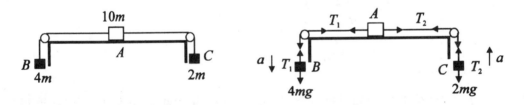

The right hand diagram shows the forces acting on the particles; there is no reason to assume the tensions in the strings are equal and so they are denoted by $T_1$ and $T_2$.  If it is assumed that $B$ has an acceleration $a$ downwards, then $A$ has an acceleration of $a$ to the left and $C$ has an acceleration $a$ upwards.  The equations of motion of the particles $A$, $B$, $C$ are respectively

$$4mg - T_1 = 4ma,$$
$$T_1 - T_2 = 10\,ma,$$
$$T_2 - 2mg = 2ma.$$

Adding these gives $a = \dfrac{g}{8}$ and substituting in the last equation gives $T_2 = 2.25\,mg$.

## Example 5.12

Find the acceleration for the configuration of Example 5.11 when the table is rough with coefficient of sliding friction 0.1.

The only difference is that there will be a frictional force acting on $A$. The reaction of the table is $10\,mg$ and the friction force is therefore $mg$ and it acts to the right since $A$ moves to the left. The middle equation then becomes

$$T_1 - T_2 - mg = 10\,ma,$$

adding the equations now gives $a = \dfrac{g}{16}$.

## Exercises 5.3

Questions 1 to 3 refer to a light inextensible string passing over a small smooth pulley, with particles of masses $m_1$ and $m_2$ attached one at each end of the string. Find, in each case, the magnitude of the acceleration of the particles and the tension in the string.

**1** $m_1 = 5$ kg, $m_2 = 3$ kg.

**2** $m_1 = 7$ kg, $m_2 = 3$ kg.

**3** $m_1 = 4M$, $m_2 = 6M$.

Questions 4 to 6 refer to a particle A of mass $m_1$ on a horizontal plane and attached by a light inextensible string, which passes over a small smooth pulley at the edge of the plane, to a particle $B$ of mass $m_2$ which hangs freely with the string vertical. The vertical plane through $B$ and the pulley contains that part of the string between $A$ and the pulley.

**4** The plane is smooth, $m_1 = 3$ kg, $m_2 = 5$ kg. Find the magnitude of the acceleration of the particles and the tension in the string.

**5** The plane is smooth, $m_1 = 1$ kg, $m_2 = 4$ kg. Find the magnitude of the force exerted on the pulley.

**6** The plane is rough with coefficient of friction 0.5, $m_1 = 3$ kg, $m_2 = 6$ kg. Find the tension in the string.

Questions 7 to 10 refer to a particle $A$ of mass $m_1$ on a plane inclined at an angle $\alpha$ to the horizontal. It is connected by a light inextensible string, which passes over a small smooth pulley at $B$ and which hangs freely with the string $BC$ vertical, to a particle $C$ of mass $m_2$. The portion $AB$ of the string is parallel to a line of greatest slope of the plane.

**7** The plane is smooth with $\alpha = 30°$, $m_1 = 1$ kg, $m_2 = 4$ kg. Find the magnitude of the acceleration of the particles.

**8** The plane is smooth with $\sin \alpha = \frac{4}{5}$, $m_1 = 5$ kg, $m_2 = 3$ kg. Find the tension in the string.

**9** The plane is rough with coefflcient of friction 0.25, $\sin \alpha = \frac{3}{5}$, $m_1 = 2$ kg, $m_2 = 8$ kg. Find the magnitude of the acceleration of the particles.

**10** The plane is rough with coefficient of friction 0.25, $\sin \alpha = \frac{5}{13}$, $m_1 = 13$ kg, $m_2 = 15$ kg. Find the magnitude of the acceleration of the particles.

## 5.4 Problems involving vehicle motion

The problems already encountered involving vehicle motion have been such that the driving force has been given explicitly. This is not what happens in practice since what is normally known for a vehicle engine is its rate of working or power (these terms will be explained in detail in 6.6). For the time being in order to look at practical problems all that you need to assume is that there is a number, the power, associated with a given engine. This is measured in watts W, with a thousand watts being a kilowatt (kW). These are also used in domestic electricity consumption.

For cars the old-fashioned unit of power, which is still often referred to, is horse power (H.P) and 1 H.P = 745.7 watt. If an engine is working at $P$ watts and the point of contact of the wheels with the ground is moving at speed $v$ ms$^{-1}$ then the driving force $F$ N at the wheels satisfies the equation

$$P = Fv.$$

The derivation and logic behind this equation will be explained in 6.6 but, for now, it will be used to work out the driving force for vehicles working at a rate of $P$ W moving with speed $v$ ms$^{-1}$. (When a car is not skidding the speed of the point of contact of the wheels is equal to that of the car and this will normally be assumed.)

Problems where the power developed by a vehicle is given are very similar to previous problems involving vehicle motion except that the force has to be worked out from the power and speed. Some problems require finding the steady speed at which a vehicle can travel when the engine is working at a given power. Since the vehicle is moving at a steady speed the acceleration is given, it is in fact zero. Sometimes the maximum speed may be given. A maximum is a stationary point so $\frac{dv}{dt} = 0$, i.e. the acceleration will be zero.

## Example 5.13

A particular motor cycle develops a maximum power of 24 kW and, when working at this rate, its speed is 40 ms$^{-1}$. Find the driving force.

The driving force $F$ N is such that
$$24000 = 40\,F,$$
giving $F$ as 600.

## Example 5.14

A car travels at constant speed, with its engine working at a rate of 40 kW, against a resistance of 1600 N. Find the speed.

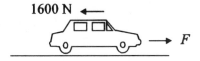

The forces acting are shown in the diagram. If $v$ ms$^{-1}$ denotes the speed of the car then $F = \dfrac{40\,000}{v}$ N. Since the acceleration is zero the force is equal to the resistance i.e.
$$\frac{40000}{v} = 1600,$$
so that $v = 25$.

## Example 5.15

The engine of a lorry of total mass 2 tonnes is working at 50 kW. The lorry is travelling at a constant speed of 20 ms$^{-1}$ along a level road. Find the total resistance to the motion.
If the power is increased to 60kW, find the acceleration of the lorry at the instant it is moving with speed 20 ms$^{-1}$ assuming that the resistances to motion remain constant.
When the power is 50kW, the driving force $F$ N is given by
$$F \times 20 = 50\,000,$$
so that $F = 2500$ N.
Since the lorry is moving at constant speed the acceleration must be zero, i.e the total force is zero so the resistance is 2500 N.
When the power is 60 kW, the driving force $F$ N at the instant the speed is 20 ms$^{-1}$ is
$$\frac{60\,000}{20}\ \text{N} = 3000\ \text{N}.$$
The total force on the lorry is $(3000 - 2500)$ N $= 500$ N, the equation of motion is
$$2000a = 500,$$
where $a$ ms$^{-2}$ is the acceleration, therefore $a = 0.25$.

## Example 5.16

A car of mass 1000 kg has a maximum speed of 35 ms⁻¹ on a level road against a resistance of 400 N. Find, assuming the engine works at the same rate and that the resistance is unchanged, its maximum speed up a hill inclined at an angle to the horizontal where $\sin \alpha = \frac{1}{7}$.

In this case the power is not given and will be assumed to be $P$ W, the driving force is $\frac{P}{35}$ N and, since at maximum speed the acceleration, and hence the total force, is zero

$$\frac{P}{35} = 400.$$

Therefore $P = 14\,000$.

The forces acting on the car when moving at speed $v$ ms⁻¹ up the hill are shown in the diagram.

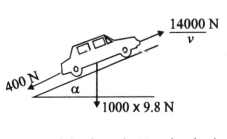

The total component of the force in N acting in the sense up the hill is

$$\frac{14000}{v} - 400 - 1000 \times 9.8 \times \sin \alpha = \frac{14000}{v} - 1800.$$

When the speed is a maximum the total force is zero i.e.

$$v = \frac{14000}{1800} = 7.9.$$

## Exercises 5.4

**1** Find the power developed when a force of 1500 N pulls a cart at a constant speed of 6 ms⁻¹.

**2** Find the power that is developed by the engine of a car moving at a speed of 12 ms⁻¹, given that the driving force is 750 N.

**3** A car, whose engine is working at a rate of 5 kW, is moving with speed 10 ms⁻¹, find the driving force.

**4** A car travels along a horizontal road against a resistance of 600 N. Given that the engine is working at a constant rate of 4.8 kW find the maximum speed of the car.

**5** A car of mass 1000 kg is moving on a horizontal road against a resistance of 600 N with the engine working at a rate of 8 kW. Find the acceleration of the car at the instant it is moving with a speed of 10 ms⁻¹.

## Dynamics of Rectilinear Motion

**6** A car of mass 1200 kg travelling along a horizontal road, with its engine working at a constant rate, against a resistance of 500 N has a maximum speed of 25 ms$^{-1}$. Find the rate at which the engine is working.

Find also the maximum speed with which the car can climb a hill inclined at an angle $\alpha$ to the horizontal, where $\sin \alpha = \frac{1}{14}$, assuming that the resistances and the rate of working of the engine are unchanged.

**7** A train of mass 300 tonnes travels along a straight level track. The resistance to motion is 18 kN. Find the tractive force required to produce an acceleration of 0.1 ms$^{-2}$, and the power in kW which is then developed by the engine when the speed of the train is 10 ms$^{-1}$.

Find also the maximum speed attainable when the engine is working at a rate of 360 kW.

**8** A car of mass 1500 kg travels up a slope of inclination $\alpha$ to the horizontal, where $\sin \alpha = \frac{1}{49}$, against constant frictional resistances of 3600 N. Find the maximum speed of the car given that the engine works at a rate of 80 kW. After reaching the top of the slope the power is switched off and the car descends a slope of inclination $\beta$ to the horizontal against the same constant frictional resistances at constant speed. Calculate $\sin \beta$.

**9** A car of mass 1000 kg has a maximum speed of 15 ms$^{-1}$, against a constant frictional force equal to one eighth of the weight of the car, up a slope inclined at an angle $\alpha$ to the horizontal where $\sin \alpha = \frac{1}{7}$. Find the maximum speed of the car on a horizontal road assuming that the engine works at the same rate.

If the car descends the same slope with its engine working at half this rate, find the acceleration of the car at the moment when its speed is 25 ms$^{-1}$.

**10** The resistive forces opposing the motion of a train, of total mass 50 tonnes are 5000 N.
(a) Find the power necessary to keep the train moving along a straight level track at a constant speed of 10 ms$^{-1}$.
(b) If this power is suddenly increased by 10 kW when the train is moving along the level track at 10 ms$^{-1}$, find the initial acceleration of the train.
(c) When the train climbs a hill, of inclination $\alpha$ to the horizontal, at a constant speed of 8 ms$^{-1}$, the engine of the train is working at a rate of 180 kW. Find the value of $\sin \alpha$.

**11** A car of mass 1600 kg climbs a slope of inclination $\alpha$ to the horizontal, where $\sin \alpha = \frac{1}{14}$, at a steady speed of 12 ms$^{-1}$. Given that the frictional resistance is 400 N, calculate the power, in kW, developed by the car.

When the car reaches the top of the slope the power is switched off and the car descends a slope of inclination $\beta$ to the horizontal, where $\sin \beta = \frac{1}{28}$. Assuming that the frictional resistance remains at 400 N, calculate the acceleration with which the car descends this slope.

12  A car of mass 1600 kg is moving along a horizontal road. The resistance to the motion of the car is 800 N. Calculate the accleration of the car at the instant when its speed is 7.5 ms$^{-1}$ and its engine is working at 15 kW.

13  When a car is moving with speed $v$ ms$^{-1}$ the resistance to its motion is $(200 + 2.5v)$ N. Find the maximum speed of the car when its engine is working at a rate of 5 kW.

## 5.5 Forces dependent on time

In practical situations forces acting will not be constant. For example even for a car working at a constant rate the force will depend on velocity, and it is often not very straightforward to find the position and velocity given the force. You have already seen that finding the displacement from the force is not difficult when the force is constant. The other case which can be done fairly easily is that when the force is known in terms of $t$. The equation of motion $F = ma$ gives the acceleration in terms of $t$ and therefore $x$ and $v$ can be found exactly as in 4.5.

### Example 5.17

A particle of mass 0.4 kg is moving under the action of a force in the positive $x$-direction which at time $t$ s is $4 \exp \dfrac{t}{4}$ N and which acts for $0 \le t \le 4$. At time $t = 0$ the particle is at rest at the point $x = 0$. Find its velocity and displacement when $t = 4$.

If $v$ ms$^{-1}$ denotes the velocity in the positive $x$-direction then Newton's second law gives

$$0.4 \frac{dv}{dt} = 4 \exp \frac{t}{4}.$$

Integrating this equation from $t = 0$ to $t = t$ gives

$$v = 160 \left( \exp \frac{t}{4} - 1 \right),$$

where the condition $v = 0$ at $t = 0$ has been used. Substituting $t = 4$ gives the velocity at $t = 4$ as 274.9 ms$^{-1}$.

Therefore 
$$\frac{dx}{dt} = 160 \left( \exp \frac{t}{4} - 1 \right),$$

where $x$ m denotes the displacement at time $t$ s, and integrating this equation from $t = 0$ to $t = t$ gives

*Dynamics of Rectilinear Motion*

$$x = 160\left(4\exp\frac{t}{4} - t\right),$$

where the condition $x = 0$ at $t = 0$ has been used. Substituting $t = 4$ gives the displacement at $t = 4$ as 1.1 km.

**Exercises 5.5**

**1** A particle of mass $m$ kg moves in a straight line under the action of a force acting along the same straight line and which at time $t$ s is $m(2 + 6t)$ N. The particle is moving at 20 ms$^{-1}$ when $t = 2$. Calculate the speed of the particle when $t = 0$.

**2** A body of mass 2 kg starts from rest at $O$ and moves along the x-axis under the action of a force $(6t - t^2)$ N acting in the positive $x$ direction. What is the speed of the body (i) after 3 s, (ii) after 9 s, from the start?

**3** A particle of mass $m$ kg starts from rest at the origin and moves in a straight line under the action of a force along this line, which at time $t$ s is $0.2\, m\, e^{2t}$ N. Find the velocity of the particle when $t = 3$, and the distance of the particle from the origin when $t = 2$.

**4** A particle moves up a line of greatest slope of a smooth inclined plane, the angle $\alpha$ made by this line of greatest slope with the horizontal being such that $\tan\alpha = \frac{3}{4}$.

There is a force acting on the particle up this line of greatest slope which at time $t$ s is given by $m(12 - 3t)$ N. Find the velocity acquired when starting from rest in (i) 2 s, (ii) $t$ s. Find the distance travelled in 3 s from rest.

## 5.6 The concept of mass

As mentioned in 5.1 the mass of a body is a fundamental property of a body and the unit of mass can be chosen independently of those of length and time. In this section the idea of mass and its measurement is described in slightly more detail.

If, for example, a small body is suspended from a light spring and set in motion vertically then, in principle, its acceleration can be measured for various extensions of the spring. If this process is repeated with a different body then its acceleration for various extensions can also be measured and this can be done for many bodies. It would be found that the accelerations at the same extension would be different, the force exerted by the spring, being dependent only on the extension, would be the same in all cases. It would also be found out that for all extensions, i.e. for different values of the force acting, the ratios of the magnitudes of the accelerations would be constant for any pair of bodies. Therefore, there exists an independent relative property of the bodies which is demonstrated by this ratio. This property is referred to as the mass of a body.

Having recognised the existence of mass, the next step is to quantify it. The accelerations produced by the same force (i.e. at the same extension) acting on different particles $P$, $Q$ and $R$ are measured and the magnitude of these accelerations are denoted by $a_P$, $a_Q$ and $a_R$ respectively. The masses of $P$, $Q$ and $R$, denoted by $m_P$, $m_Q$ $m_R$ are then defined so that

$$\frac{m_Q}{m_P} = \frac{a_P}{a_Q}, \quad \frac{m_R}{m_P} = \frac{a_P}{a_R}.$$

If one of these particles ($P$ say) is then chosen to have unit mass, the masses of the other particles can be found by measuring the accelerations. It can also be verified experimentally that, for mass defined in this way, the mass of the combined particle $Q$ and $R$ is the sum of the masses of $Q$ and $R$.

Therefore, in principle a method of measuring mass can be established, and the definition of the unit of mass is independent of the choice of units of length, time and force.

## Miscellaneous Exercises 5

**1** A tug, of mass 7000 kg, pulls a boat, of mass 4000 kg, by means of an inextensible horizontal tow rope along a straight canal. The resistive forces opposing the motions of the tug and the boat are 1400 N and 900 N respectively. Find the tension in the rope when the tug is accelerating at 1 ms$^{-2}$ and the driving force exerted by the tug.

**2** When a car of mass 1800 kg is moving with speed 20 ms$^{-1}$ on a straight horizontal road its engine is switched off. The car is brought to rest in a distance of 500 m by a constant retarding force $F$ newtons. Find $F$ and the time taken for the car to come to rest.

**3** The speed of a car of mass 600 kg, moving along a level road, is reduced from 18 ms$^{-1}$ to 8 ms$^{-1}$ by a constant retarding force of 1800 N. Find the time taken for the car to reduce speed and the distance travelled

**4** A lift, starting from rest, moves with constant acceleration for 4 s and with constant velocity for the next 8 s. A constant retardation is then applied so that the lift comes to rest in 4 s at a height of 6 m above its starting point. There is a packing case of mass 20 kg on the floor of the lift.

Find   (i) the acceleration and retardation of the lift,

      (ii) the reaction of the floor of the lift on the case during each stage of the motion during the first stage of the motion.

**5** A man of mass 60 kg stands on the floor of a lift which descends with acceleration 0.6 ms$^{-2}$. Find the total force exerted between the floor of the lift and the man's feet. A man stands on a weighing machine which rests on the floor of a stationary lift, and the dial of the machine shows 75 kg. The lift then moves upwards and the dial then shows a constant value of 80 kg. Find the upward acceleration of the lift.

The lift then starts to slow down and the dial reading changes to a constant value of 70 kilograms. Find the retardation of the lift.

**6** A heavy particle is suspended by a spring balance from the ceiling of a lift. When the lift moves up with constant acceleration $a$ ms$^{-2}$ the balance shows a reading 1.8 kg. When the lift descends with constant acceleration $3a$ ms$^{-2}$ the balance shows a reading 1.2 kg. Find the mass of the particle and the value of $a$.

**7** A particle of mass $m$ is moving vertically upwards with speed $v$ at the instant it enters a fixed horizontal layer of material of thickness $a$ which resists its motion with a constant force of magnitude $R$. Find the condition to be satisfied by $v$ so that the particle passes through the layer.

**8** A lady of mass 49 kg stands in a lift. Find the reaction of the floor of the lift on the lady when the lift is

(a) moving down with constant acceleration 0.2 ms$^{-2}$,

(b) moving up with acceleration 0.3 ms$^{-2}$.

Another lady of the same mass stands in the lift when it is stationary. She has to support herself by a walking stick which she presses on the floor of the lift. Assuming that the force exerted by the walking stick on the floor of the lift is vertical, and of magnitude 98 N find the reaction of the floor of the lift on the lady.

**9**(a)

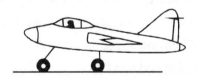

An aeroplane of mass 7000 kg lands at 50 ms$^{-1}$. It is brought to rest by braking and resistive forces which are assumed to be constant and of total magnitude 7 kN.

(i) Find the retardation of the aeroplane.

(ii) Determine the distance that the aeroplane travels before coming to rest.

(b)

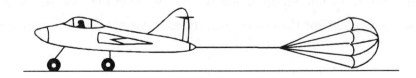

By fitting a tail parachute, with the additional equipment being of mass 200 kg, the distance that the aeroplane travels before coming to rest is reduced by 500 m.

(i) Find, assuming it to be constant, the retardation of the plane when the parachute is used.

(ii) Determine the total drag force in this case.

**10** A particle slides with acceleration 3 ms$^{-2}$ down a line of greatest slope of a rough plane inclined at an angle $\alpha$ to the horizontal, where $\tan \alpha = \frac{3}{4}$. Calculate the coefficient of friction between the particle and the inclined plane.

**11** A car of mass 1000 kg, whose engine is working at a rate of $P$ watts, moves at a constant speed of 20 ms$^{-1}$ on a horizontal road. Find, in terms of $P$, the total frictional resistance on the car.

The car then freewheels (i.e. without the engine exerting any force) down a hill inclined at an angle $\alpha$ to the horizontal, where $\sin \alpha = \frac{1}{14}$, at constant speed. Find $P$.

Assuming the same total frictional resistance and that the engine is working at the rate of $P$ watts, find the numerical value of the acceleration of the car at the instant it is moving with speed 7 ms$^{-1}$.

**12** A car of mass 1000 kg moves, with its engine working at its maximum rate, at a constant speed of 10 ms$^{-1}$ up a hill inclined at an angle $\alpha$ to the horizontal, where $\sin \alpha = \frac{1}{14}$. The frictional resistance to motion is $R$ newtons. Express the maximum rate of working in terms of $R$.

With the engine still working at its maximum rate the car descends the hill at a constant speed of 20 ms$^{-1}$. Given that the frictional resistance is now $4R$ newtons, find $R$.

**13** A locomotive of mass $M$ kg working at a rate of $R$ kW ascends a straight track which is inclined at an angle $\alpha$ to the horizontal. When the speed is $v$ ms$^{-1}$ the acceleration is $a$ ms$^{-2}$. Find an expression for the resistance at speed $v$ ms$^{-1}$.

**14** A car of mass 1500 kg is moving along a horizontal road. The resistance to the motion of the car is 750 N. Assuming that the car's engine works at 15 kW find

(i) the maximum constant speed at which the car can travel,

(ii) the acceleration of the car when its speed is 8 ms$^{-1}$.

**15** A car of mass 1300 kg tows a trailer of mass $m$ kg along a level road. The resistance on the car is 1000 N and that on the trailer is $1.5m$ N. Find the total power developed by the engine when $m = 600$ and the car and trailer are moving at a steady speed of 20 ms$^{-1}$.

For $m = 700$ and the car moving at a speed of 25 ms$^{-1}$ and its engine working at a steady rate of 75 kW find

(i) the acceleration of the car and trailer,

(ii) the tension in the coupling between the car and trailer.

State whether or not the car can maintain a steady speed of 25 ms$^{-1}$ with the engine working at a steady rate of 25 kW.

Given that the safest minimum speed for travel on a motorway is 18 ms$^{-1}$ and that the car engine is working at a rate of 72 kW find the range of possible values of $m$ so that the car and trailer can travel at steady speeds which are no lower than the safest minimum speed.

**16** A car travels at a constant speed of 20 ms$^{-1}$ on a horizontal road against constant resistance of 1000 N. Find the rate of working of the engine.

The car is then attached to a caravan by a towbar and the resistance to the motion of the caravan is 800 N. Given that the rate of working of the engine is 35 kW, find the maximum speed of the car and caravan on a horizontal road and obtain the tension in the tow bar.

The total mass of the car and the caravan is 1.8 tonnes and the car pulls the caravan directly up a hill which is inclined at an angle of $\alpha$ to the horizontal where $\sin \alpha = \dfrac{1}{35}$.

Given that the car is working at a rate of 42 kW, calculate the acceleration up the hill when the car and caravan are travelling at 14 ms$^{-1}$. The non-gravitational resistances to the motion of the car and the caravan may be assumed constant.

**17** The maximum speed of a car, whose engine can develop 15 kW, on a level road is 30 ms$^{-1}$. Find the total resistance to the motion of the car at its maximum speed.

Given that the non-gravitational resistance to the motion of the car varies as the square of the speed and that the mass of the car is 800 kg, determine the power developed by the engine when the car moves at a constant speed of 35 ms$^{-1}$ directly up a hill which is inclined at an angle $\alpha$ to the horizontal where $\sin \alpha = \dfrac{1}{21}$.

**18** A car of mass 800 kg is moving at a constant speed of 50 ms$^{-1}$ on a level road. The non-gravitational resistance to the motion of the car at all speeds and on all roads is 700 N. Calculate the rate at which the engine of the car is working.

When the car climbs directly up a certain hill, it has a maximum speed of 25 ms$^{-1}$ and the engine is working at a rate of 20 kW. Calculate the angle of inclination of the road to the horizontal.

Find the acceleration of this car when travelling at 25 ms$^{-1}$ up a hill inclined at an angle $\alpha$ to the horizontal, where $\sin \alpha = \dfrac{1}{21}$, with its engine working at the rate of 15 kW.

**19** A car of mass 800 kg is pulling a trailer of mass 200 kg up a hill inclined at an angle $\alpha$ to the horizontal, where $\sin \alpha = \frac{1}{14}$. When the total force exerted by the engine is 1000 N the car and trailer move up the hill at a steady speed. Find the total frictional resistance to the motion of the car and trailer.

When the car and trailer are travelling at a steady speed of 10 ms$^{-1}$ up the hill the power exerted by the engine is instantaneously increased by 2 kW. Find

(i) the instantaneous acceleration,

(ii) the instantaneous tension in the coupling between the car and trailer, given that the total resistance on the trailer is 75 N.

**20** Two particles of mass $M$ and $4M$ respectively are connected by a light inextensible string passing over a smooth fixed pulley. The particles are released from rest with the hanging parts of the string vertical. Find

(i) the acceleration of the particles,

(ii) the force exerted by the string on the pulley in the subsequent motion.

**21** A lift when empty is of mass 1200 kg and when it moves upwards with an acceleration of 1 ms$^{-2}$ and carrying $N$ passengers each of mass 80 kg, the tension in the cable is $\frac{1}{2} T$ newtons. The same tension occurs when $N + 5$ people, each of mass 80 kg, are accelerating downwards at 1 ms$^{-2}$. Find $N$ and $T$.

**22** Two particles of mass $4m$ and $m$ respectively are joined by a <u>light inextensible string passing over a smooth peg</u>. The particles are held at rest with the string taut and then released from rest.

(a) State which of the underlined words enables you to assume that

    (i) the tension is constant in the string on either side of the peg,

    (ii) the tension is the same on both sides of the peg at the points of contact with the string.

(b) Write down the equation of motion of each particle and determine the acceleration of the particles.

**23** The diagram shows a particle $P$ of mass $m$ on a horizontal table and attached to a particle $Q$ of mass $5m$ by a light inextensible string passing over a smooth pulley at the edge of the table. The string is perpendicular to the straight edge of the table and $Q$ can move in a vertical line. The particles are held at rest with the string taut and then released,

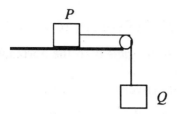

(a) Assuming that the table is smooth, show that the speed of $Q$ after it has dropped a distance $d$ from rest is $\sqrt{\dfrac{5gd}{3}}$.

(b) In an actual experiment with the particles it is found that the speed of $Q$ after dropping a distance $d$ is less than $\sqrt{\dfrac{5gd}{3}}$. Suggest one reason for the decrease in speed found by experiment.

**24**

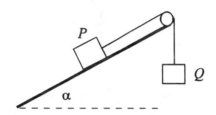

The diagram shows a particle $P$ of mass $5m$ on a rough plane inclined at an angle $\alpha$ to the horizontal where $\sin \alpha = \dfrac{3}{5}$. The coefficient of friction between $P$ and the plane is $\dfrac{1}{4}$. A light inextensible string attached to the particle passes parallel to a line of greatest slope of the plane and over a small smooth pulley at the top of the plane. A particle $Q$ of mass $10m$ is attached to the other end of the string and $Q$ can move freely in a vertical line. Given that the particles are released from rest at time $t = 0$ find the tension in the string. After $Q$ has dropped a distance $10a$ it hits a horizontal plane and then stays at rest on the plane. Find the further distance travelled by $P$ before first coming to instantaneous rest.

**25**

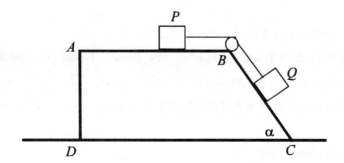

The diagram shows a vertical section $ABCD$ of a block of wood fixed on a horizontal plane. $AB$ is horizontal and $BC$ is inclined at an angle $\alpha$ to the horizontal where $\sin \alpha = \dfrac{4}{5}$. Particles $P$ and $Q$, of mass $m$ and $5m$ respectively, are placed on $AB$ and $BC$ and joined together by a light inextensible string passing over a smooth pulley at $B$. The particles are then released from rest. Find, assuming that $P$ does not reach $B$ and $Q$ does not reach $C$, the acceleration of the particles and the tension in the string when

(i) $AB$ and $BC$ are smooth,

(ii) $AB$ and $BC$ are both equally rough, the coefficient of friction being $\dfrac{1}{3}$.

**26**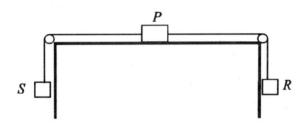

The diagram shows a particle $P$ of mass $2m$ on a rough horizontal table attached by light inextensible strings to particles $R$ and $S$ of mass $6m$ and $2m$ respectively. The coefficient of friction between $P$ and the table is 0.5. The strings pass over light smooth pulleys on opposite sides of the table so that $R$ and $S$ can move freely with the strings perpendicular to the table edges. Given that the system is released from rest with the strings taut find the magnitude of the common acceleration of the particles and the tension in the string joining $P$ and $S$.

After falling a distance $d$ from rest the particle $R$ strikes the floor which is such that $R$ is brought to rest and then remains at rest. Find the further distance that $S$ rises.

**27** A particle $P$ of mass 0.2 kg is acted on by a variable force so that its velocity in ms$^{-1}$ at time $t$ s is $16 - t^2$. Find the distance covered by $P$ from time $t = 0$ until it comes to rest instantaneously.

Find also the force acting on the particle at time $t$ s.

**28** Two bodies of mass 1 kg and 2 kg, initially at the points $A$ and $B$ respectively, start from rest at $t = 0$ and move along the horizontal straight line $AB$. The first is acted on by a force of $(6t + 2)$ newtons towards $B$ and the second by a force of $2(12t^2 + 16)$ newtons towards $A$. Find

(a) the speed of each body after 1 second

(b) the distances covered by each body during the first second.

Given that the bodies collide after 1 second, find the distance $AB$.

## Dynamics of Rectilinear Motion

**29** An electric train of mass $M$ kg moves from rest along a straight level track. The tractive force of the motor, initially $P$ N, decreases at a constant rate with time to $R$ N over a period of $T$ s and then remains constant at $R$ N. The total resistance to motion is $R$ N. Show that the acceleration $a$ of the train at time $t$ s after it starts to move is given, for $0 \le t \le T$, by

$$Ma = P + (R - P)\frac{t}{T} - R.$$

Find the maximum speed achieved by the train and the distance it travels before reaching that speed.

Find the power developed by the motor at time (i) $\dfrac{T}{2}$, (ii) $\dfrac{3T}{2}$.

**30** A car of mass 1000 kg moves along a horizontal road with acceleration proportional to the cube root of the time $t$ seconds after starting from rest. When $t = 27$, the speed of the car is 8 ms$^{-1}$. Find the rate at which the engine driving the car is working when $t = 64$. Frictional resistances may be neglected.

# Chapter 6

# Work, Energy and Power

After working through this chapter you should
- be able to calculate the work done by constant forces and those dependendent only on position,
- know what is meant by kinetic energy, potential energy and power,
- be able to use the work-energy principle to find the work done by a force,
- know when the total mechanical energy is conserved and use the principle of conservation of mechanical energy to solve simple problems,
- be able to calculate the power necessary for engines such as water pumps to carry out their tasks.

## 6.1 Work done by a constant force

You will probably connect the word work with something which requires effort, for example if you push a wheelbarrow up a hill, or just cycle up a hill, you will feel that you have done some work. In Mechanics it is possible to give a precise definition of work which is, in fact, consistent with this general idea of expending effort. The simplest definition is for the case when the force is constant.

The work done by a constant force moving a particle along a line is the product of the distance moved by the particle and the component of the force in the direction of motion.
The work done by a force of 1 newton moving through 1 metre is the unit of work and is called the joule ( J).
The work done can be positive or negative according as to whether the force is acting towards, or away from, the direction in which the particle is moved.

## Example 6.1

Find the work done by a horizontal force of magnitude 30 N which pushes a heavy parcel a distance of 4 m along a smooth floor.

The work is the product of distance and the component of force in the direction of motion and is $30 \times 4$ J $= 120$ J.

## Example 6.2

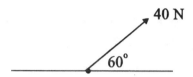

The diagram shows a particle on a horizontal wire being pulled along by a constant force of magnitude 40 N inclined at an angle of $60°$ to the wire. Find the work done by the force in moving the particle a distance of 0.4 m along the wire.

The component of force along the wire is $40 \cos 60°$ N $= 20$ N and therefore the work done is $20 \times 0.4$ J $= 8$ J.

Sometimes it is not the work done by a force that is required but the work done against it and you have to be clear as to which you are calculating. For example if a parcel of mass $m$ is lifted a vertical distance $h$ then the work done by gravity is $-mgh$ since the component of gravity in the direction of the motion of the particle is $-mg$. In order to find the work that has to be done in the lifting it is assumed that the lifting is carried out extremely slowly at a constant speed so that the total force acting is zero and the lifting force therefore exactly balances the force of gravity. The magnitude of the lifting force is therefore $mg$ and the work done by the lifting force is $mgh$. For a wheelbarrow moving up a hill the force of friction and the component of the weight along the hill are both in the opposite direction to the motion and so the work done by them is negative. If the wheelbarrow is moving at constant speed then the pushing force just balances the other forces at all times and the work done by it is minus the work done by friction and gravity. In general the work done against a particular force is taken to be minus the work done by the force.

**Example 6.3**

A particle of mass 0.5 kg is pulled a distance of 4 m, at constant speed, up a slope inclined at an angle $\alpha$ to the horizontal where $\tan \alpha = \frac{3}{4}$. The pulling force acts parallel to a line of greatest slope of the plane. The coefficient of friction between the particle and the plane is 0.5. Find the work done in this motion by the pulling force.

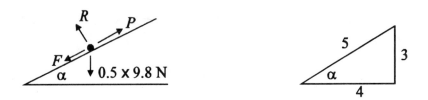

The left hand diagram diagram shows the forces acting on the particle with $F$, $R$ and $P$ denoting the magnitudes of the friction force, the normal reaction and the pulling force respectively. The reaction of the plane does no work since its component in the direction of motion is zero, similarly it is only the component of the weight along the plane that does work. There are two ways of carrying out the calculation. One way is to find the work done by the friction and the work done by gravity separately and add them together. The work done by the pulling force is then minus this work. The other way is to say, since the particle is moving at constant speed and has therefore no acceleration, that the force acting along the plane is equal in magnitude but opposite in direction to the components of friction and gravity along the plane and work out the work done by this force. This second method is the one that will be used.

The reaction is found by resolving perpendicular to the plane and is $0.5 \times 9.8 \times \cos \alpha$ N. The right hand diagram above shows that $\cos \alpha = \frac{4}{5}$ and therefore the reaction is 3.92 N and the force of friction is 1.96 N. The component of the weight down the plane is $0.5 \times 9.8 \times \sin \alpha = 2.94$ N. The total pulling force is therefore $(2.94 + 1.96)$ N $= 4.9$ N. The work done is therefore $4.9 \times 4$ J $= 19.6$ J.

Before calculating work done you should check that the force is constant before multiplying it by the distance moved. This method only makes sense for constant force. For forces which vary with position you need to use the definition in the following section.

## Exercises 6.1

**1** Find the work done by a horizontal force of magnitude 2 N pushing a particle a distance of 0.8 m along a horizontal surface.

**2** A climber of mass 55 kg, climbing at constant speed, does 1000 J of work. Find the distance climbed.

**3** A parcel of mass 5 kg is lifted, at constant speed, through a height of 5 m. Find, modelling the parcel as a particle, the work done against gravity.

**4** A particle of mass 0.2 kg is pulled at a constant speed of 5 ms$^{-1}$ along a rough horizontal surface. Given that the coefficient of friction is 0.3 find the work done against friction in 4 s.

**5** A particle of mass 0.3 kg is pulled at a constant speed up a smooth plane inclined at an angle of 20° to the horizontal. Find the work done against gravity in moving the particle a distance of 3 m up the plane.

**6** A packing case is pulled a distance of 4 m at constant speed across a horizontal floor by a rope attached to it and inclined at an angle of 40° to the horizontal. Given that the total work done by the pulling force is 80 J find the tension in the rope.

**7** A toboggan of mass 20 kg is pulled at a constant speed a distance of 12 m up a snowy slope inclined at an angle of 20° to the horizontal. The coefficient of friction between the toboggan and the snow is 0.6. Find

(i) the work done against gravity,

(ii) the work done against friction.

Given that the toboggan is being pulled by a rope attached to its front find the tension in the rope for the two cases when

(a) the rope is parallel to the slope,

(b) the rope is at an angle of 45° to the slope.

**8** A particle of mass 5 kg is pulled with constant speed up a plane inclined at an angle $\alpha$ to the horizontal where $\sin \alpha = \frac{3}{5}$. The coefficient of friction between the plane and the particle is $\frac{1}{3}$. Given that the pulling force acts along a line of greatest slope find the work done in moving the particle a distance of 10 m.

**9** A man and bicycle are of total mass 100 kg. He travels, at constant speed, a distance of 1 km up a hill inclined at an angle $\alpha$ to the horizontal where $\sin \alpha = \frac{1}{20}$. The other resistances acting on him and directly opposing his motion total 20 N. Find the total work done by the cyclist.

**10** The cyclist in the previous question travels at a constant speed down a hill inclined at an angle $\alpha$ to the horizontal where $\sin \alpha = \dfrac{1}{160}$ and does 1200 J of work in travelling a distance of 400 m. Find, assuming that it is constant, the total resistance to his motion.

## 6.2 Work done by a force dependent only on position

If a force is not constant but depends on position then it is not sensible to define work as force times distance since you would not know at what point to calculate the force! In order to find a more sensible definition it is necessary to have a graphical interpretation of work.

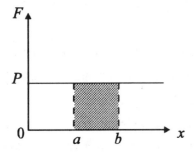 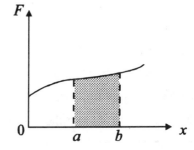

If the component of force in the positive $x$ direction is denoted by $F$ then the left hand diagram shows the graph of $F$ for a constant force of magnitude $P$ and the right hand diagram a possible graph when $F$ is not constant but depends on position i.e. $x$. If $F$ has the constant value $P$ then the work done moving from $x = a$ to $x = b$ is $P(b - a)$; this is the area under the straight line in the left hand diagram. Therefore a sensible definition of the work done moving from $x = a$ to $x = b$ when $F$ depends on $x$ would be the area under the curve in the right hand diagram between the lines $x = a$ and $x = b$. This is the definition used, and since the area under a curve can be represented as an integral, the work done in a displacement of the point of application of the force from $x = a$ to $x = b$ is defined as

$$\int_a^b F(x)\,dx.$$

This definition, as has been shown above, reduces to the original definition when the force is constant.

The integral definition is actually valid for all forces but, when the force depends on other variables such as $t$ and $v$, the integral is not easy to evaluate.

## Example 6.4

When the displacement of a particle from the origin is $x$ m the force acting on it is of magnitude $(10 + 4e^{-x})$ N and acts in the positive $x$ direction. Find the work done when the particle moves from $x = 0$ to $x = 1$.

The work done is

$$\int_0^1 (10 + 4e^{-x})\, dx \text{ J} = 10 - 4(e^{-1} - 1) \text{ J} = 14 - 4e^{-1} \text{ J}.$$

## Exercises 6.2

**1** When a particle $P$ is at a point on the positive $x$-axis at a distance of $x$ m from the origin the force in the positive $x$ direction is of magnitude $4x$ N. Find the work done by the force when
(i) the particle moves from the point $x = 1$ to $x = 2$,
(ii) the particle moves from the point $x = 4$ to $x = 3$.

**2** When the displacement of a particle from the origin is $x$ m the force in the positive $x$ direction acting on it is denoted by $F(x)$ N. Find the work done in moving the particle from the point $x = a$ to the point $x = b$ when
(i) $F = 4x^3 + 3x^2$, $a = 1$, $b = 2$,
(ii) $F = 3 + 4e^{-x}$, $a = 0$, $b = 1$,
(iii) $F = \dfrac{4}{x^2}$, $a = 2$, $b = 4$.

**3** A truck is pulled along a straight horizontal track by a horizontal force whose magnitude, when the truck is at a distance of $x$ m from its starting point, is $(8 - \dfrac{x}{20})$ N. Find the work done as the truck moves a distance of 100 m.

## 6.3 Work done by the tension in an elastic string

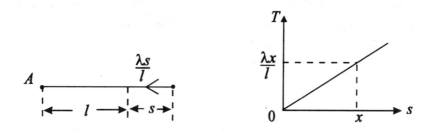

If an elastic string has one end fixed at a point $A$ and the other end is extended a distance $s$ beyond its natural length then Hooke's law shows that the tension $T$ is of magnitude $\dfrac{\lambda s}{l}$ where $\lambda$ is the elastic modulus and $l$ is its natural length. This tension acts away from the extended end as shown in the left hand diagram and therefore its component towards the fixed end is $\dfrac{\lambda s}{l}$. The right hand diagram shows the variation of tension with extension $s$. The magnitude of the work done by the tension when the string is extended from $s = 0$ to $s = x$ is therefore the area under the straight line between $s = 0$ and $s = x$; this is $\dfrac{\lambda x^2}{2l}$. Since the tension is directed away from the direction of extension the work done by it in the extension is $-\dfrac{\lambda x^2}{2l}$. The same result is obtained by integrating the force in the direction of the extension, i.e. $-\dfrac{\lambda s}{l}$, from $s = 0$ and $s = x$, this gives

$$-\int_0^x \frac{\lambda s}{l}\,ds = -\frac{\lambda x^2}{2l}.$$

The particular letter used for the variable of integration is not very important, it is what is known as a dummy variable, but it is better, and avoids confusion, not to use as a variable of integration a variable which occurs in one of the limits.

Therefore the work done against the tension in extending an elastic string a distance $x$ is $\dfrac{\lambda x^2}{2l}$ and this can also be shown, in a similar way, to be true when a spring is compressed a distance $x$.

The work done against the tension in increasing the extension of an elastic spring from $a$ to $b$ is

$$\int_a^b \frac{\lambda s}{l}\,ds = \frac{\lambda}{2l}(b^2 - a^2).$$

## Exercises 6.3

Find the work done in extending an elastic string of natural length $l$ m and modulus $\lambda$ N a distance of $x$ m from its unstretched length when

**1** $l = 2$, $\lambda = 100$, $x = 0.2$.
**2** $l = 3$, $\lambda = 300$, $x = 0.4$.
**3** $l = 2.5$, $\lambda = 500$, $x = \dfrac{l}{4}$.

Find the work done in increasing the extension of an elastic string of natural length $l$ m and modulus $\lambda$ N from $a$ m to $b$ m when

**4** $l = 2$, $\lambda = 300$, $a = 0.2$, $b = 0.4$.
**5** $l = 3$, $\lambda = 600$, $a = 0.4$, $b = 0.7$.

**6** A light elastic string of modulus 30 N and natural length 1 m hangs unstretched. Find the work done by a man in stretching the string a distance of 0.1 m. How much extra work has he to do to extend the string by a further 0.3 m?

**7** An elastic string of natural length 1 m and modulus 50 N is horizontal and has one end fixed and is stretched by applying a force of 5 N to the other. Find the work done by this force.

## 6.4 Kinetic energy and the work-energy principle

The word energy occurs very often in everyday life and there are many different forms of energy e.g. heat energy, electrical energy, chemical energy. The use of the word is often very loose but in any scientific work it is necessary to be a little more precise. The energy of a particle is its capability to do work. This is not a particularly clear definition but one which you may understand a little bit more clearly by considering one particular form of energy encountered in Mechanics. This is **Kinetic Energy (K.E.)**, which, for a particle of mass $m$ moving with speed $v$, is defined as $\frac{1}{2} mv^2$. Kinetic energy is effectively the energy possessed by a particle by virtue of its motion. The unit of kinetic energy is the joule (J).

The kinetic energy can be related to the capacity of a particle to do work by imagining a particle of mass $m$ set off with speed $v$ on a rough horizontal floor. The friction force will act in the opposite direction to the motion and reduce speed. The point of application of the force of friction moves and therefore the force does work until the particle comes to rest. Therefore the moving body had some capacity to do work. This can be made more precise by assuming that the friction force $F$ is constant so that the retardation of the particle is $-\dfrac{F}{m}$. Denoting the distance that the particle has moved by $s$ and applying $v^2 = u^2 + 2as$ with $v = 0$, $u = v$ and $a = -\dfrac{F}{m}$ gives

$$0 = v^2 - 2\dfrac{F}{m}s,$$

and therefore $Fs = \dfrac{1}{2}mv^2$. Therefore the work done in reducing the speed to zero is equal to the particle's original kinetic energy. This should makes clearer the idea of energy as a capability to do work.

The above result relating the work done to a change in kinetic energy is a simple example of a general result proved in 6.7 which is known as the **work-energy principle** and which states that

### Change in K.E = Total work done by the forces acting.

This general principle can be used to solve many mechanical problems very simply. A particular advantage of using it is that there are some forces which do no work and they do not have to be considered at all. The simplest example of such forces is the reaction of a smooth surface, if a particle moves along such a surface then the reaction has no component in the direction of motion and therefore does no work. Another example is the tension in a taut light string.

The tension act away from the ends as shown in the diagram, and if $A$ moves a distance $d$ then so does $B$ and therefore the total work done is zero. This is still true if the string passes over a smooth pulley.

Another advantage of using the work-energy principle is that, even if the precise nature of the forces acting is not known, it can be used to find the total work done provided the change in kinetic energy is known.

Many problems involving motion under constant forces can often be solved just as easily by using the constant acceleration formulae and it often is just a matter of preference which method you use. In examinations you may however be told something like "by energy considerations find ... ", and then you would have to use the work-energy principle. The principle of conservation of mechanical energy described in 6.5 is a very slight variant of the work-energy principle and in many circumstances is effectively identical with it.

## Example 6.5

A particle is dropped from rest at a height of 3 m above a horizontal floor, find the speed with which it hits the floor.

The mass of the particle is not given and it will therefore be denoted by $m$ kg.

If the speed with which the particle hits the floor is $v$ ms$^{-1}$ then the change in its kinetic energy is $\frac{1}{2}mv^2$ J. The force of gravity is acting in the direction of motion so the work done by it is $3 \times 9.8\,m$ J = $29.4\,m$ J. Equating these gives $v^2 = 58.8$ and $v = 7.67$.

This problem can also be solved by using the constant acceleration formulae.

The downwards acceleration is 9.8 ms$^{-1}$ and substituting $a = 9.8$, $s = 3$, $u = 0$ in $v^2 = u^2 + 2as$ gives the same answer.

## Example 6.6

A particle of mass 0.3 kg, dropped from a height of 4 m reaches a speed of 8 ms$^{-1}$ just as it hits the floor. Find the work done by air resistance

The change in kinetic energy is $\frac{1}{2} \times 0.3 \times 64$ J = 9.6 J. The work done by gravity is $0.3 \times 4 \times 9.8$ J = 11.16 J. The work-energy principle then gives

$$9.6 = 11.16 + \text{work done by air resistance},$$

the work done by the air resistance is therefore $-1.56$ J, the minus sign showing that the resistance is acting in the opposite direction to the motion.

## Example 6.7

A particle of mass 0.3 kg moves on a smooth horizontal plane under the action of a horizontal force of magnitude 12 N. Find the speed of the particle after it has moved a distance of 5 m from rest.

The reaction of the plane does no work and neither does the force of gravity since it also is perpendicular to the motion. (These forces need not have to be considered separately since, as mentioned in 5.1, forces perpendicular to the line of motion are in equilibrium. Therefore the nett force is zero and therefore so is the work done.)

The work done during the motion is therefore $12 \times 5$ J = 60 J. This is equal to the change in kinetic energy, i.e. $\frac{1}{2} 0.3 \times v^2$ J, where the final speed is $v$ ms$^{-1}$. This gives

$$v = \sqrt{400} = 20.$$

The acceleration along the plane of the particle is $\frac{12}{0.3}$ ms$^{-2}$ = 40 ms$^{-2}$ and again using $v^2 = u^2 + 2as$ with this value of $a$ gives the same result.

## Example 6.8

A heavy parcel of mass 10 kg is pushed along a rough horizontal floor by a force and its speed increases from 1 ms$^{-1}$ to 2 ms$^{-1}$ whilst it travels a distance of 5 m. Given that the coefficient of friction is 0.5 find the work done by the pushing force.

Find also this force for the two cases

(i) when it is assumed to be constant,

(ii) when it is assumed to be of the form $kx$ where $x$ is the distance moved from the point where the speed is 1 ms$^{-1}$.

The change in kinetic energy is $5 \times (2^2 - 1^2)$ J = 15 J. The reaction is 98 N and therefore the force of friction is 49 N. The force of friction acts in the opposite direction to the motion and therefore the work done by it is $-49 \times 5$ J = $-245$ J.

The work done by gravity and by the normal reaction are, as in Example 6.7, zero.

Therefore the work-energy principle gives

$$15 \text{ J} = \text{Work done by pushing force} - 245 \text{ J},$$

the work done by the pushing force is therefore 260 J.

If the pushing force is assumed to be $F$ N then the work done would be $5F$ J so $F = 52$.

If the force is assumed to be of the form $kx$ N, where $x$ m is the displacement from the point where the speed is 1 ms$^{-1}$, then the work done is $\int_0^5 kx \, dx$ J = 12.5 $k$. This gives

$k = 20.8$.

## Example 6.9

A particle of mass 0.3 kg moves along the x-axis under the action of an attractive force directed towards the origin and of magnitude $\dfrac{6}{x^2}$ N when the particle is at a distance of $x$ m from the origin. It is projected in the positive $x$ direction with a speed of 10 ms$^{-1}$ from the point where $x = 1$. Find its speed for $x = 2$.

Denoting the speed when $x = 1$ by $v$ ms$^{-1}$ the change in kinetic energy is $(0.15v^2 - 0.15 \times 100)$ J. The force is acting in the opposite direction to the motion so the work done is $-\int_1^2 \dfrac{6}{x^2} dx$ J $= \left(\dfrac{6}{x}\right)_{x=1}^{x=2}$ J $= -3$ J.

The work energy principle gives

$$0.15v^2 - 15 = -3,$$

giving

$$v = \sqrt{80} = 8.94.$$

## Example 6.10

When the displacement from the origin of a particle of mass 0.4 kg is $x$ m the component of force acting on it in the positive $x$ direction is $(8 + 6e^{-x})$ N. Given that the particle has speed 2 ms$^{-1}$ at the origin find its speed when $x = 1$.

If the speed when the particle is at a distance of 1 m is denoted by $v$ ms$^{-1}$ the change of kinetic energy is $(0.2 v^2 - 0.2 \times 2^2)$ J $= (0.2 v^2 - 0.8)$ J.
The motion is in the direction of the force and the work done moving from $x = 0$ to $x = 1$ is $\int_0^1 (8 + 6e^{-x})dx$ N. The work-energy principle gives

$$0.2 v^2 - 0.8 = \int_0^1 (8 + 6e^{-x})dx,$$

i.e. $$0.2 v^2 = \int_0^1 (8 + 6e^{-x})dx + 0.8 = 14.8 - 6e^{-1} = 12.59,$$

giving $v = 7.93$ ms$^{-1}$.

For problems where the forces acting are dependent only on position it is possible, from the work-energy principle, to derive the principle of conservation of mechanical energy. This is slightly simpler to apply in cases where all the forces are of a standard type.

## Exercises 6.4

**1** Find the kinetic energy of
(i) a particle of mass 0.2 kg moving with a speed of 8 ms$^{-1}$,
(ii) a car of mass 800 kg moving at a speed of 22 ms$^{-1}$,
(iii) a woman of mass 55 kg running at a speed of 4 ms$^{-1}$.

**2** A car of mass 600 kg decreases its speed from 25 ms$^{-1}$ to 20 ms$^{-1}$. Find the change in its kinetic energy.

**3** A particle of mass 0.4 kg moving with speed 20 ms$^{-1}$ has its kinetic energy suddenly reduced by 10 J. Find the new speed of the particle.

**4** A bullet of mass 0.02 kg moving with speed 100 ms$^{-1}$ enters a block of wood and comes to rest after moving a distance of 0.05 m. Find the resistance of the wood
(i) assuming that it is constant,
(ii) assuming that it is directly proportional to the distance the bullet has entered the wood.

**5** When the brakes are applied a car of mass 1000 kg comes to rest, from a speed of 12 ms$^{-1}$, in a distance of 40 m. Assuming that the other resistances acting on the car are of magnitude 90 N find the work done by the braking force.

**6** A car of mass 1000 kg is pushed by two men, one at each back corner. Each one exerts a force of 100 N at an angle of 15° to the direction of motion of the car and there is a constant resistance of magnitude 80 N acting. Find

(i) the work done by the men in moving the car a distance of 10 m,

(ii) the speed attained from rest in 10 m.

**7** A boy of mass 30 kg slides a vertical height of 5 m from rest down a water chute. Find the speed he attains.

**8** A man pushes a wheel barrow, of mass 35 kg, with a force of magnitude 20 N at an angle of 30° to the horizontal. There is a resistance to motion of magnitude 10 N. Find, given that the wheel barrow is moved a distance of 10 m from rest,

(i) the work done by the man,

(ii) the speed attained.

**9** A ball of mass 0.4 kg thrown vertically upwards with a speed of 10 ms$^{-1}$ comes to instantaneous rest at a height of 4.5 m above the point of projection. Find the magnitude of the work done by the air resistance. Assuming that the magnitude of the work done by the air resistance is the same on the downwards path, find the speed with which the ball returns to the point of projection.

**10** The speed of a particle of mass 0.15 kg sliding on a rough floor decreases from 7 ms$^{-1}$ to 4 ms$^{-1}$ while it moves a distance of 6 m. Find

(i) the work done by friction,

(ii) the coefficient of friction.

**11** A particle of mass 0.4 kg is projected with speed 15 ms$^{-1}$ up a rough plane inclined at an angle of 40° to the horizontal, the coefficient of friction being 0.3. Use the work-energy principle to find the distance that the particle moves up the plane before coming to rest.

**12** A particle of mass 0.5 kg, free to move along the $x$-axis, is attracted towards the origin by a force $\dfrac{10}{x^3}$ N when the particle is at a distance of $x$ m from the origin. Given that it is projected from the point $x = 1$ with speed 10 ms$^{-1}$ in the positive $x$ direction find its speed when $x = 2$.

**13** A particle of mass 0.5 kg moves along the $x$-axis under the action of an attractive force directed towards the origin and of magnitude $\dfrac{10}{x^2}$ N when the particle is at a distance of $x$ m from the origin. It is projected in the positive $x$ direction with a speed of $u$ ms$^{-1}$ from the point where $x = 1$. Find the condition that $u$ has to satisfy in order that the particle does not return to its initial position.

*Work, Energy and Power*

## 6.5 Potential energy and energy conservation

For forces, like those considered in 6.2, which depend only on position it is possible to define a second form of energy, **the potential energy**, which can be used to solve relatively complicated problems where more than one type of such force is involved. This approach is particularly useful where two or more standard type forces, such as gravity and the force in an elastic string, are involved.

**Potential Energy (P.E.)** can only be defined for forces which depend on position and is the work done against the force to move its point of application (usually a particle) from a standard position to its present position. Alternatively it can be regarded as the work done by the force in moving its point of application from its present position to a standard one. Potential energy is energy possessed by a particle by virtue of its position.

If the force in the positive $x$ direction is $F(x)$ the work done against it in moving from the standard position $x = a$ to the position $x = x$ is

$$-\int_a^x F(s)\,ds,$$

where the standard position (i.e. the position of zero potential energy) is $x = a$.

**Potential energy due to gravity**

For the force of gravity the standard position is often taken to be ground level, but any appropriate level is satisfactory. If $x$ is measured upwards then the force in the $x$ direction acting on a particle of mass $m$ is $-mg$ so the potential energy due to gravity (the gravitational potential energy) at height $h$ above the zero level is $mgh$. Similarly the potential energy at height $H$ below the zero level is $-mgH$.

If a particle is released from rest at a height $h$ above ground it immediately starts moving i.e. work is being done. This again is consistent with energy being a capacity to do work. By the time the particle reaches the ground its potential energy is zero but it will have gained kinetic energy.

In working out the gravitational potential energy of a particle you should remember that it is positive if the particle is above the zero level and negative if it is below this level.

**Potential energy of an elastic string**

For an elastic string the zero of potential energy is the point where the string is just unstretched i.e. the extension $s$ is zero. The work done against the tension in extending the string moving from $s = 0$ to $s = x$ is $\int_0^x \frac{\lambda s}{l} \, ds = \frac{\lambda x^2}{2l}$, where $\lambda$ is the elastic modulus and

$l$ is the natural length. The potential energy of a stretched string, extended a distance $x$, is therefore $\int_0^x \frac{\lambda s}{l} \, ds = \frac{\lambda x^2}{2l}$.

This is sometimes called the elastic potential energy or the energy stored in the string. The same result holds for a spring compressed by a distance $x$.

The potential energy is effectively a method of calculating the work done by particular kinds of forces and the idea of potential energy can be used to state the work-energy principle in a slightly different way. The details are given in 6.7 where it is shown that

**Total mechanical energy is constant provided all forces are dependent only on position,**

where the total mechanical energy is the sum of the kinetic energy and the potential energy of all the forces acting.

This is the principle of conservation of mechanical energy and forces which depend only on position are, for obvious reasons, called conservative forces.

Forces which do no work, like those normal to the direction of motion and the tension in a taut string, can be ignored in working out potential energy.

If there are non-conservative forces present the principle of conservation of mechanical energy has to be replaced by

**Change in total mechanical energy = Work done by non-conservative forces**

If a force is conservative, i.e. depends only on position, the work done in moving from $x = a$ to $x = b$ is $\int_a^b F(x)\,dx$ and the work done in going back from $x = b$ to $x = a$ is $\int_b^a F(x)\,dx = -\int_a^b F(x)\,dx$. The total work done from $x = a$ to $x = b$ and back from $x = b$ to $x = a$ is therefore zero. Therefore another way of testing whether a force is conservative is to say that the total work from a point to another one and back is zero. This test can be used to show that frictional forces, which at first sight appear to be constant, are not conservative. If a particle slides on a rough horizontal plane, then there is a frictional force of magnitude $\mu mg$ acting on the particle, where $\mu$ is the coeffficient of friction. Moving directly from $x = 0$ to $x = a$ the work done by the force of friction is $-\mu mga$, since the motion is in the opposite direction to the force. On moving back from $x = a$ to $x = 0$ a further amount of work $-\mu mga$ is done by the friction force since it is again acting in the opposite direction to the motion. Therefore a non-zero amount of work is done moving from a point and then back. Therefore frictional forces are not conservative.

## Work, Energy and Power

The principle of conservation of mechanical energy can be used whenever the forces are conservative but it is most useful in problems involving springs or strings and the force of gravity since the potential energies for these are known. Questions involving the use of conservation of energy will give information at various points of the motion and it is a good idea to set out a solution in a way which shows clearly the energies at these points and then apply the conservation of mechanical energy for all possible points.

### Example 6.11

A particle of mass $m$ is thrown vertically upwards with an initial speed $u$. Find (i) the greatest height reached, (ii) the speed when the particle is at a height $h$ which is less than the greatest height.

The points involved in the problem are the initial point $O$, the point $A$ at height $h$ and the point $B$ of greatest height which will be taken to be at a height $H$. At this point the speed will be zero. The speed at $A$ will be denoted by $v$.

At $O$,  P.E. $= 0$,  K.E. $= \frac{1}{2}mu^2$,

At $A$,  P.E. $= mgh$,  K.E. $= \frac{1}{2}mv^2$,

At $B$,  P.E. $= mgH$,  K.E. $= 0$.

Putting the total mechanical energy at $O$ equal to that at $B$ gives

$$\frac{1}{2}mu^2 = mgH,$$

so that $H = \dfrac{u^2}{2g}$. Equating the total mechanical energy at $A$ to that at $O$ gives

$$\frac{1}{2}mv^2 + mgh = \frac{1}{2}mu^2.$$

Solving for $v$ gives $\qquad v = \sqrt{u^2 - 2gh}.$

This particular problem could have been solved just as simply by using the equations of motion under uniform acceleration.

## Example 6.12

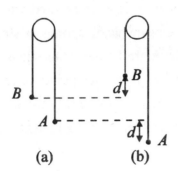

Diagram (a) shows two particles $A$ and $B$ of masses $5m$ and $m$ attached one to each end of a light inextensible string passing over a smooth light pulley. Initially the particles are held at rest with the string taut and then released. Find, by using conservation of energy, the speed of $A$ when it has dropped a distance $d$.

The total work done by the tension is zero and therefore, if both particles are considered, the total mechanical energy is conserved. Both particles also move with the same speed and if one drops a distance $d$ the other rises by the same distance. The initial position and that when particle $A$ has dropped a distance $d$ are shown in diagrams (a) and (b).

Diagram (a)          Particle $A$ : K.E. = 0, P.E. = 0,
                                 Particle $B$ : K.E. = 0, P.E. = 0.

Diagram (b),          Particle $A$ : K.E. = $\frac{1}{2} 5mv^2$, P.E. = $-5mgd$,
                                 Particle $B$ : K.E. = $\frac{1}{2} mv^2$, P.E. = $mgd$.

Equating the total energy at both positions gives
$$3mv^2 - 4mgd = 0.$$
Therefore
$$v = \sqrt{\frac{4gd}{3}}.$$

This example could have been solved as in 5.3 by finding the acceleration and using the constant acceleration formulae but, in this instance, the use of conservation of energy avoids the calculation of acceleration.

## Example 6.13

A particle is released from rest at a point $A$ on a smooth plane inclined at an angle of $30°$ to the horizontal. Find its speed when it has moved to a point $B$ a distance of 8 m down the plane.

*Work, Energy and Power*

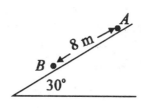

Apart from gravity the only force acting on the particle is the reaction of the plane. This is perpendicular to the motion and does no work; therefore total mechanical energy is conserved.

The mass is not given so it will be denoted by $m$ kg and the speed at $B$ will be denoted by $v$ ms$^{-1}$. At $A$ the particle will be at a height of $8 \times \sin 30° $ m $= 4$ m above $B$.

At $A$ :  K.E. $= 0$,  P.E. $= m9.8 \times 4$ J $= 39.2m$ J.

At $B$ :  K.E. $= \frac{1}{2} mv^2$ J,  P.E. $= 0$.

Equating the total energies gives

$$39.2m = \frac{1}{2} mv^2,$$

so that $v^2 = 78.4$ and $v = 8.85$.

This problem could have been solved by finding the component along the plane of the force of gravity; this is $\frac{1}{2} mg$. The acceleration down the plane is therefore $\frac{1}{2} g$ and the constant acceleration formulae can be used.

### Example 6.14

A particle of mass 0.4 kg is attached to one end of an elastic string of modulus 2 N and natural length 0.25 m. The other end of the spring is attached to a fixed point $O$ on a smooth horizontal table. The string is extended a distance of 0.05 m and the particle then released from rest. Find

(i) the speed of the particle when the string returns to its unstretched position,
(ii) the speed when the extension is 0.02 m.

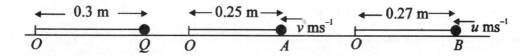

The diagrams show the initial position, when there is zero extension and when the extension is 0.02 m.

There are three points that have to be considered the initial point $Q$, the point $A$ of zero extension and the point $B$ with extension 0.02 m. The velocities at $A$ and $B$ are denoted by $v$ ms$^{-1}$ and $u$ ms$^{-1}$, respectively. The nett force perpendicular to the table is zero and therefore does no work.

The formula for the potential energy of an elastic string shows that when the string is extended by $x$ m its potential energy is $\dfrac{\lambda x^2}{l} = \dfrac{2x^2}{2 \times 0.25}$ J $= 4x^2$ J.

At $Q$,          P.E. $= 4 \times (0.05)^2 = 0.01$ J,   K.E. $= 0$.

At $A$,          P.E. $= 0$,          K.E. $= \dfrac{1}{2} \times 0.4v^2 = 0.2v^2$ J,

At $B$,          P.E. $= 4 \times (0.02)^2$ J,     K.E. $= 0.2u^2$ J.

Equating the energy at $Q$ to that at $A$ gives
$$0.2 v^2 = 0.01,$$
so
$$v = 0.22.$$
Equating the energy at $Q$ to that at $B$ gives
$$0.2 u^2 + 4\times(.02)^2 = 0.01,$$
so that
$$u = 0.2.$$

### Example 6.15

A particle of mass 0.6 kg is attached to one end of an elastic string of modulus 352.8 N and of natural length 0.5 m. The other end of the string is attached to a fixed point $O$. Initially the particle is held at $O$ and released from rest.

Find the maximum extension of the string. This is a model of bungee jumping.

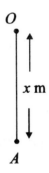

The two positions of the particle are shown in the diagram. It will drop until it comes to instantaneous rest at the point $A$ at a distance $x$ m below $O$, so the kinetic energy is zero at the start and the end of the motion.

There are two terms which contribute to the potential energy, the gravitational potential energy and the elastic potential energy. The formula for the potential energy of the elastic string shows that the elastic potential energy when the particle is at a depth of $x$ m below $O$ is $\dfrac{352.8(x-0.5)^2}{2 \times 0.5}$ J provided $x > 0.5$, otherwise it is zero.

At $O$,   Gravitational P.E. = 0,
Elastic P.E. = 0,
K.E. = 0.

At $A$,   Gravitational P.E. = $-0.6 \times 9.8\, x = -5.88x$ J,
Elastic P.E. = $\dfrac{352.8(x-0.5)^2}{2 \times 0.5}$ J,
K.E. = 0.

Equating the total energy at $O$ and $A$ gives
$$\dfrac{352.8(x-0.5)^2}{2 \times 0.5} - 5.88x = 0.$$
Expanding $(x-0.5)^2$ and collecting the tems in the equation gives
$$352.8x^2 - 358.68x + 88.2 = 0.$$
This is a quadratic equation for $x$. Using the formula for solving the quadratic gives the roots as 0.6 and 0.417. The equation obtained is only valid when the elastic string is taut, i.e. $x > 0.5$, so the correct solution is $x = 0.6$.

## **Example 6.16**

Find the speed of the particle in the above example when the particle is
(i) at a depth 0.3 m below $O$,
(ii) at a depth 0.55 m below $O$.

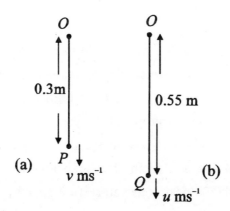

Diagram (a) shows the point $P$ corresponding to the first case and diagram (b) shows the point $Q$ corresponding to the second case. The velocities at $P$ and $Q$ are denoted by $v$ ms$^{-1}$ and $u$ ms$^{-1}$ respectively.

## Case (i)

The string will be slack at $P$ and therefore will have no elastic potential energy. The point $O$ is taken to be the zero level of potential energy.

At $O$,    Gravitational P.E. = 0,
Elastic P.E. = 0,
K.E. = 0.

At $P$,    Gravitational P.E. = $-0.6 \times 9.8 \times 0.3$ J = $-1.764$ J,
Elastic P.E. = 0,
K.E. = $\frac{1}{2} \times 0.6 \, v^2$ J = $0.3v^2$ J.

Equating the total energies at $O$ and $P$ gives
$$0.3v^2 - 1.764 = 0,$$
so that    $v = 2.42$.

## Case (ii)

At $Q$,    Gravitational P.E. = $-0.6 \times 9.8 \times 0.55$ J = $-3.234$ J,
Elastic P.E. = $\dfrac{352.8(0.55-0.5)^2}{2 \times 0.5}$ J = $0.882$ J,
K.E. = $0.3u^2$

Equating the total energies at $O$ and $Q$ gives
$$0.3u^2 - 3.234 + 0.882 = 0.$$
The solution of this is    $u = 2.8$.

## Example 6.17

A particle of mass 0.4 kg is attached to one end of an elastic string of natural length 1 m and modulus 19.6 N, the other end of the string being attached to a fixed point $O$. Initially the particle is in equilibrium with the string vertical. It is then pulled to a point $A$ at a distance of 0.2 m below the equilibrium position and released from rest. Find

(i) the length of the string in the equilibrium position,
(ii) the speed of the particle when it is at a depth of 0.1 m below the equilibrium position,
(iii) the position of the particle when it next comes to rest.

The first step is to find the equilibrium point $E$. If the extension of the string is $y$ m then
$$\frac{19.6y}{1} = 0.4 \times 9.8,$$
so that $y = 0.2$ and so the length of the string in the equilibrium position is 1.2 m.

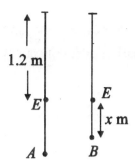

The diagram shows the initial point $A$ of the particle and that when the particle is at a point $B$ at a depth of $x$ m below $E$. The extension in this position is $(0.2 + x)$ m and therefore the elastic potential energy is $\dfrac{19.6(0.2 + x)^2}{2}$. The initial elastic potential energy is found by substituting $x = 0.2$ in this. The zero of potential energy is taken at the equilibrium position.

At $A$,  Gravitational P.E. $= -0.4 \times 9.8 \times 0.2$ J $= 0.784$ J

Elastic P.E. $= \dfrac{19.6(0.4)^2}{2}$ J $= 1.568$ J

K.E. $= 0$.

At $B$,  Gravitational P.E. $= -0.4 \times 9.8 x$ J $= -3.92x$ J

Elastic P.E. $= \dfrac{19.6(0.2 + x)^2}{2}$ J

K.E. $= 0.2v^2$ J.

Equating the total energies at $A$ and $B$ gives

$$\dfrac{19.6(0.2 + x)^2}{2} - 3.92x + 0.2v^2 = 1.568 - 7.84.$$

The equation simplifies to $\quad 9.8x^2 + 0.2v^2 = 0.392$.

Substituting $x = 0.1$ gives $v = 1.21$.

Substituting $v = 0$ gives $x = \pm 0.2$ so the particle next comes to rest at a height of 0.2 m above the equilibrium position where the string just becomes slack. Since this point is not an equilibrium point the total force there is not zero and therefore it will move away from it. You can check that the force at this point is directed downwards so the particle will move down. It will stop instantaneously at the point $A$ and the cycle then repeats itself.

### Example 6.18

Answer (ii) and (iii) of Example 6.17 when the particle is pulled a distance of 0.3 m below the equilibrium point.

The only difference between this question and the previous one is that the initial position corresponds to $x = 0.3$. This means that the equation of energy is now
$$\frac{19.6(0.2 + x)^2}{2} - 0.4 \times 9.8\, x + 0.2v^2 = \frac{19.6(0.5)^2}{2} - 0.4 \times 9.8 \times 0.3.$$
Expanding $(0.2 + x)^2$ and collecting terms gives
$$9.8x^2 + 0.2v^2 = 0.882.$$
substituting $x = 0.1$ gives $\quad v = 1.98$.

Substituting $v = 0$ gives $x = \pm 0.3$ so the particle next comes to rest at a height of 0.3 m above the equilibrium point. The string will however have become slack when the particle is at a distance of 0.2 m above the equilibrium position and therefore the above equation will only be valid for $x \le 0.2$. At this position $0.2v^2 = 0.49$. From this position on the only potential energy is that due to gravity and taking the zero of potential energy at this point gives, on applying conservation of energy $0.49 = 0.4 \times 9.8\, h$, where $h$ denotes the height when the speed is zero and this is 0.125. So the height above the equilibrium position is 0.3125 m.

## Exercises 6.5

**1** Find the change in potential energy in the following cases
(i) a particle of mass 0.3 kg moved upwards through a vertical distance of 1.5 m,
(ii) a man of mass 60 kg walking a distance of 50 m along and down a hill inclined at an angle of $45°$ to the horizontal,
(iii) a stone of mass 2 kg dropped a distance of 5 m from a bridge.

**2** Find the increase in the gravitational potential energy of a man of mass 70 kg who climbs to the tenth floor of a block of flats. (Ground floor is floor 0). The distance between the floors is 3 m.

**3** A stone is dropped from rest from a bridge to the water 10 m below, find the speed with which it hits the water.

**4** A particle is projected with speed 5 ms$^{-1}$ up a line of greatest slope of a smooth plane inclined at an angle of $40°$ to the horizontal. Use the principle of conservation of energy to find the distance the particle moves along the plane before coming to instantaneous rest.

**5** Answer the previous question when the plane is rough with coefficient of friction 0.2.

**6** A particle of mass $m$ rests on a smooth horizontal table and is connected by a light inextensible string, passing over a smooth pulley, to a particle of mass $3m$. The particles are held at rest with the string taut and then released. Find, by using conservation of energy, the speed of the particles when the heavier one has dropped a distance $h$.

**7** Two particles of masses 0.8 kg and 0.6 kg are connected by a light inextensible string passing over a small smooth pulley. They are released from rest with the string taut. Find their speeds when the heavier particle has dropped a distance of 0.2 m.

**8**

The diagram shows a spring, with one end fixed, in a horizontal tube. The spring is of natural length 0.2 m and elastic modulus 160 N. A particle of mass 0.02 kg is placed on the free end and the spring compressed a distance of 0.05 m and then released. Find the speed of the particle when the spring is at its unstretched position.

**9** Two particles each of mass $m$ are connected by a light elastic string of natural length $a$ and modulus $4\,mg$. The string lies on a smooth horizontal table and is stretched by equal forces until it is of length $1.6\,a$. Both particles are then released simultaneously. Assuming that both particles move with the same speed find the value of the speed when they are a distance $a$ apart.

In questions 10 to 12 one end of a light elastic string, of modulus $kmg$ N and natural length $l$ m, is fixed at a point $O$ and a particle of mass $m$ kg is attached to the other end. The particle is then released from $O$. Find the extension of the string when the particle first comes to instantaneous rest.

**10** $k = 1$, $l = 1$
**11** $k = 4$, $l = 3$.
**12** $k = 5$, $l = 2$.

In questions 13 to 14 one end of a light elastic string, of modulus $kmg$ N and natural length $l$ m, is fixed at a point $O$ and a particle of mass $m$ kg is attached to the other end. The particle is initially at rest in equilibrium and is then pulled down a distance $d$ m and released from rest. Find the distance below $O$ of the point where the particle first comes to instantaneous rest.

**13** $l = 3$, $k = 6$, $d = 0.5$.
**14** $l = 2$, $k = 8$, $d = 0.4$.

## 6.6 Power

A machine, in practice, is required not only to do a certain amount of work but to do that work in a limited interval of time. A powerful car shows its power by accelerating rapidly i.e. it produces kinetic energy more rapidly than a car of lesser power.

**Power** is the rate at which work is done. If 1 joule is produced in 1 second, the rate of working is 1 watt. The watt (W) and the kilowatt (kW = $10^3$ W) are the standard units of power.

If the point of application of a force $F$ moves a small distance $\delta x$ in a small time interval $\delta t$ then the change in work, denoted by $\delta W$ is $F \, \delta x$. The rate of doing work is found by dividing $\delta W$ by $\delta t$ and letting $\delta t$ become very small. Therefore the rate of working, i.e. the power, is the value of $F \dfrac{\delta x}{\delta t}$ as both $\delta x$ and $\delta t$ become very small. This, from the idea of a derivative as a rate of change, is $F \dfrac{dx}{dt}$ and, from the definition of velocity, is $Fv$.

This was assumed in 5.4 when working out problems for moving vehicles. The rate of working of the engine of a vehicle can be calculated from analysing its motion.

It is then assumed that all this power is transmitted without loss to the driving wheels.

If the work done between some standard time and time $t$ is denoted by $W(t)$, since it could vary with time, then the definition of power $P$ gives

$$P = \frac{dW}{dt}.$$

Since $W$ depends on $t$ then so will $P$ and therefore the work done between $t = 0$ and $t = T$ is found by integrating this i.e.

$$\text{Work done} = \int_0^T P \, dt .$$

For a constant power, but only then,

Work done over a time interval $T$ is $PT$.

The relation between power and work can be used to work out the power necessary for a pump to bring water up through a height $h$ and then pump it out at speed $v$ as in the diagram.

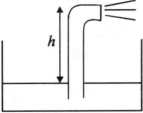

## Work, Energy and Power

The work done by the pump is converted into the kinetic and potential energy of the water and the amount of work done per second is therefore $\frac{1}{2}mv^2 + mgh$, where $m$ denotes the mass per second lifted. Since the power $P$ is assumed to be constant this work is equal to $P$ and therefore

$$P = \frac{1}{2}mv^2 + mgh.$$

### Example 6.19

The force acting on a particle of mass 0.2 kg and moving along a straight line is such that, at time $t$ s, the velocity of the particle is $t^4$ ms$^{-1}$. Find the rate at which the force acting on the particle is working.

The first step is to find the force, this is $0.2 \times$ acceleration. The acceleration can be found by differentiating the velocity and is $4t^3$ ms$^{-2}$. Therefore the rate of working is

$$0.8\, t^3 \times \text{velocity} = 0.8\, t^7 \text{ W}.$$

### Example 6.20

The rate of working at time $t$ s, of the force acting on a particle 0.4 kg moving on a straight line is $6t^5$ W. Given that the particle has speed 2 ms$^{-1}$ at time $t = 0$ find its speed at time $t = 1$.

The change in kinetic energy is the work done and the work done between $t = 0$ and $t = 1$ is

$$\int_0^1 6t^5 dt \text{ J} = 1 \text{ J}.$$

If the speed when $t = 1$ is $v$ ms$^{-1}$ then the change of kinetic energy is $0.2v^2 - 0.8$ J. Therefore $v^2 = 9$ and $v = 3$.

### Example 6.21

A pump is required to raise 100 kg of water a second through a height of 15 m and discharge it through a nozzle with speed 10 ms$^{-1}$. Find the minimum rating of pump required if the pump is
(i) 100% efficient,
(ii) 50 % efficient.

The change in P.E. per second is $100 \times 9.8 \times 15$ J = 14700 J.
The change in K.E. per second is $50 \times 10^2$ J = 5000 J.
The total change in energy per second, i.e. the work done per second is 19 700 J. Therefore the minimum rate of working, if the pump is 100% efficient is 19.7 kW. For 50% efficiency the rate would have to be 39.4 kW

## 6.7 Derivation of basic results

In this section a derivation of the work-energy principle will be given for the case of a particle of mass $m$, moving along the $x$-axis, under the action of a force whose component in the positive $x$ direction. is denoted by $F$. The velocity in the positive $x$ direction at time $t$ will be denoted by $v$.

Newton's second law of motion gives
$$m\frac{dv}{dt} = F.$$

Multiplying this equation by $v$ gives
$$mv\frac{dv}{dt} = Fv.$$

The left hand side can be written, on using the product rule for differentiation, as $\frac{d}{dt}\left(\frac{1}{2}mv^2\right)$ which is the rate of change of kinetic energy. The right hand side is the power, i.e. the rate of doing work. The equation is therefore

Rate of change of kinetic energy = Rate of doing work.

Therefore integrating this gives

Change of kinetic energy over any interval = Work done over the interval.

This is the work-energy principle in its simplest form. The proof is effectively for a particle moving on a line and the work done is that done by forces along the line. It also holds when the work done by forces not along the line is included, this is because the work done by the components of these force perpendicular to the line will be zero since the motion of these components will be perpendicular to the line.

If the only forces acting are dependent on position then introducing the potential energy produces the principle of conservation of mechanical energy. This can be shown as follows.

The work done by a force $F$ in moving from $x = x_1$ to $x = x_2$ is $\int_{x_1}^{x_2} F(x)\, dx$. This can be rewritten from the properties of integrals as
$$\int_a^{x_2} F(x)\, dx - \int_a^{x_1} F(x)\, dx.$$

The potential energy at $x = x_1$, for example, is $-\int_a^{x_1} F(x)\, dx$, therefore the above expression for the work done is equal to

$-$P.E. at $x = x_2$ $+$ P.E. at $x = x_1$.

## Work, Energy and Power

The work energy principle can therefore be rewritten as

(K.E. at $x = x_2$ − K.E. at $x = x_1$) = (− P.E. at $x = x_2$ + P.E. at $x = x_1$) .

This can again be rearranged as

(K.E. at $x = x_2$ + P.E. at $x = x_2$) = (K.E. at $x = x_1$ + P.E. at $x = x_1$).

The sum of the kinetic and potential energy is the total mechanical energy and therefore

**Total Mechanical energy is constant provided all forces are dependent only on position.**

Potential energy can only be defined for conservative forces. If there are non-conservative forces present then the work done by all the forces is − change in P.E. of the conservative forces + the work done by the non-conservative forces. The principle of conservation of mechanical energy has then to be replaced by

**Change in total mechanical energy = Work done by non-conservative forces**

### Exercises 6.6

**1** A particle is of mass 0.5 kg and the component of its velocity in the positive $x$ direction at time $t$ s is $4 \exp 3t$ ms$^{-1}$. Find the rate at which the force acting on it is working.

**2** Find the total work done in 10 s by a force when the rate of working is

(i) 3 kW,

(ii) $3\left(1 - \exp\left(-\dfrac{t}{10}\right)\right)$ kW.

**3** A particle of mass 0.8 kg has speed 4 ms$^{-1}$ at time $t = 0$ s and the force acting on it is working at a rate of $10t^9$ W at time $t$ s. Find its speed when $t = 2$.

**4** A water pump is to raise 50 kg of water a second through a height of 20 m. The water emerges as a jet with speed 50 ms$^{-1}$. Find the kinetic energy and the potential energy given to the water each second and hence find the power that the pump would have to develop if

(i) it were 100% efficient,

(ii) it were 75% efficient.

**5** A pump delivers 220 kg of water per minute, the water being delivered in a horizontal jet at a speed of 30 ms$^{-1}$. Find the kinetic energy of the water delivered each second. The efficiency of the engine driving the pump is 35%. Find the rate at which this engine is working.

## Miscellaneous Exercises 6

**1** A particle is thrown vertically upwards with a speed of 5 ms$^{-1}$ from a point 2 m above the ground. Find

(a) the greatest height above the ground reached,

(b) the speed of the particle when it hits the ground.

**2** A girl throws a ball vertically upwards so that its initial speed is 10 ms$^{-1}$.

(a) Draw a diagram showing the forces acting on the stone and state clearly the cause of each force.

(b) Find the maximum height that the ball would reach if the only force acting were that due to gravity and which may be assumed to be constant.

(c) It turns out that the ball, which has mass 0.2 kg, only reaches a height of 4.5 m; find the work done by the resistance as the ball travels from the girl's hand to its maximum height.

**3** A boy of mass 60 kg starting with a speed of 1 ms$^{-1}$ slides down a chute in a swimming pool and strikes the water at a speed of 9 ms$^{-1}$. Find the work done against friction if the chute is 5 m high.

If the length of the slide is 20 m find the frictional force, assuming it to be uniform.

**4** A toboggan run is straight, 1213 m long and drops 157 m from start to finish. One day a toboggan and its rider with a combined mass of 112 kg, starting from rest, achieved a speed of 117 kmh$^{-1}$ at the finish.

(a) Calculate the gain in kinetic energy.

(b) Find the loss in potential energy.

(c) Determine the work done against resistive forces assumed constant.

**5** A vehicle of mass 4000 kg is moving up a hill inclined at an angle $\alpha$ to the horizontal where $\sin \alpha = \frac{1}{20}$. Its initial speed is 2 ms$^{-1}$. Five seconds later it has travelled 15 m up the hill and its speed is then 6 ms$^{-1}$. Find the change in the kinetic energy and potential energy of the vehicle.

Given that the engine is working at a constant rate of 44 kW find the total work done against the resistive forces (which may not be assumed to be constant) during this five second period.

**6** An elastic spring which obeys Hooke's law has natural length 0.5 m. When the extension is 0.05 m the tension in the spring is 50 N. Find the work done when the spring is extended from a length of 0.6 m to a length of 0.7 m.

## Work, Energy and Power

**7** The gravitational force per unit mass at a distance $r$ ($> R$) from the centre of the earth is $\dfrac{gR^2}{r^2}$ where $R$ is the radius of the earth and $g$ is the acceleration due to gravity on the earth's surface. Find the work done in moving a particle of mass $m$ from $r = 2R$ to $r = 3R$. Find the speed when $r = 3R$ of a particle whose speed when $r = 2R$ was $\sqrt{\dfrac{gR}{2}}$.

**8** A catapult consists of two lengths of elastic, each of modulus 20 N and natural length 0.2 m. The catapult is stretched so that the length of each elastic is increased by 0.08 m. Ignoring the effect of gravity and using energy considerations find the speed that the catapult will give to a stone of mass 0.02 kg.

**9** The end $A$ of a light elastic string $AB$ of natural length 1.2 m is fixed. When a particle of mass 2.4 kg is attached to the string at $B$ and hangs freely under gravity the extension of the string is 0.09 m. Find the modulus of elasticity of the string.
The particle is now pulled vertically a further 0.12 m and released from rest. By energy considerations find the greatest height above the point of release in the subsequent motion.

**10** A particle $P$ of mass 0.02 kg is attached to one end of a light elastic spring of natural length 0.5 m and modulus 1.6 N, the other end of which is attached to a fixed point $A$ on a smooth horizontal table. The particle is released from rest on the table when the spring is straight and its extension is 0.25 m. Find the speed of the particle
(i) when the spring is at its natural length,
(ii) when the spring is compressed by 0.10 m.

**11** A particle of mass 2 kg is attached to two elastic strings, each of natural length 0.5 m and modulus 15 N, the other ends of the strings being attached to two fixed points $A$ and $B$ which are at a distance of 1 m apart in the same horizontal line. The particle is dropped from rest from the midpoint of $AB$.
(a) Show that the tension $T$ N in each string is given by $T = 15(\sec\theta - 1)$, where $\theta$ denotes the angle between each string and the horizontal.
(b) Find, correct to two decimal places, the acceleration of the particle when $\theta = 60°$.
(c) Use energy considerations find, correct to two decimal places, the speed of the particle when $\theta = 60°$.

**12** A particle is suspended from a fixed point $O$ by a light elastic string which is of natural length $a$ and hangs in equilibrium at a distance $\dfrac{5a}{4}$ below $O$. Given that the particle is released from rest at $O$ find the distance it falls before it first comes to rest.

## Work, Energy and Power

**13** One end $O$ of a light elastic string of natural length $4l$, is attached to a fixed point. A particle $P$ of mass $m$ is attached to the other end $P$ of the string and the string hangs in equilibrium with $OP = 5l$. The particle is pulled down vertically a further distance $\frac{l}{2}$ and released from rest. Show that $P$ rises a distance $l$ before first coming to instantaneous rest.

**14** Find also, for the configuration in the previous question, the maximum height to which $P$ would rise if it had been released from rest at a depth of $2l$ below the equilibrium position.

**15** One end of a light elastic string of modulus 4.9 N and natural length 0.5 m is attached to a fixed point $A$ and a particle of mass 0.1 kg is attached to the other end. The particle is held at $A$ and released from rest. Its speed when it has dropped a distance of $x$ m is $v$ ms$^{-1}$.
(i) Write down an expression for its speed when $x \leq 0.5$.
(ii) Show by use of energy that, for $x \geq 0.5$, $v^2 = 117.6\ x - 9.8x^2 - 24.5$.

**16** In the dangerous sport of bungee diving an individual attaches one end of an elastic rope to a fixed point on a river bridge. He/she is then attached to the other end and jumps over the bridge so as to fall vertically downwards towards the water. The rope should be such that the diver comes to rest just above the surface of the water. In order to find which particular ropes are suitable, experiments are carried out with weights, rather than people, attached to the rope. In one experiment it was found that when a weight of mass $m$ was attached to a particular rope of natural length $a$ and dropped from a bridge at a height $3a$ above the water level then the weight just reached the level of the water. Show that the modulus of elasticity of the rope is $\frac{3mg}{2}$.

The weight of mass $m$ is removed and a weight of mass $\frac{5m}{2}$ is then attached to this rope and dropped from the same height. Find the speed of the weight just as it reaches the water.

When the weight emerges from the water its speed has been reduced to zero by the resistance of the water. Show, by using conservation of energy, and assuming that the rope does not slacken, that the subsequent speed $v$ of the weight at height $h$ above the water level is given by

$$v^2 = \frac{gh}{5a}(2a - 3h).$$

Describe the subsequent motion of the weight.

**17** A particle of mass *m* is attached to one end of an elastic string, of modulus 4*mg* and natural length *a*, the other end of the string being attached to a fixed point *O*. The particle is released from a point at a distance $\frac{5a}{3}$ directly below *O*.

Find

(i) the height to which the particle will rise,

(ii) the speed of the particle when it is at a distance of $\frac{3a}{2}$ below *O*.

**18** A water pump raises 40 kg of water a second through a height of 20 m and ejects it with a speed of 45 ms$^{-1}$. Find the kinetic energy and potential energy per second given to the water and the effective rate at which the pump is working.

**19** A Venetian blind consists of a fixed top bar and ten moveable bars each of mass 0.05 kg and of negligible thickness. When the blind is fully extended the distance between consecutive bars is 0.05 m. Find the work done in closing it.

# Chapter 7

# Impulse and momentum

After reading this chapter you should
- know what is meant by impulse and momentum,
- know and be able to apply the impulse momentum principle,
- know and be able to apply the principle of conservation of momentum to problems of collision of inelastic bodies,
- know Newton's elastic law and be able to apply it, together with the principle of conservation of momentum, to collisions of elastic bodies.

## 7.1 Impulse momentum principle

There are many examples in everyday life where the velocity of a body is changed fairly quickly. One example is when the brakes of a car are applied very suddenly, another is when a tennis ball is hit. In all these case a force is acting on a body for a very short period of time. Since the time that the force acts is very short it is actually very difficult to measure either the time or the force. It turns out however that it is a quantity depending on both the force and the time that actually determines the change in velocity. This quantity is known as the impulse and the idea of impulse is possibly most easily understood by considering the case of a constant force acting for a short period of time. This is not a good model for sharp blows but the basic definition will then be extended to give a more realistic model.

It will be assumed that the component of a force in a given direction is denoted by $P$ and that it acts on a particle of mass $m$ for time $T$. During this period it is assumed that the velocity of the particle in the given direction changes from $u$ to $v$. The acceleration due to the force is $\frac{P}{m}$ and applying $v = u + at$ gives

## Impulse and momentum

$$v - u = \frac{P}{m} T.$$

This can be rearranged slightly as

$$m(v - u) = PT.$$

The right hand side of this equation is directly proportional to the change in velocity and is dependent only on the force and the time for which it acts. This is known as the **impulse of the force** which, for a constant force, is defined as the product of the component of the force in a given direction and the time for which it acts. The unit of impulse is the newton second (Ns).

The product of the mass of a particle and its velocity is known as its **momentum** (the unit of momentum is therefore also the newton second) and therefore the above equation is equivalent to

$$\text{change in momentum} = \text{impulse}.$$

This is the impulse momentum equation; so far it has only been shown to hold for constant forces but it is true whatever force is acting. The momentum as defined is actually the linear momentum but as you will not meet any other type of momentum in your course there will be no confusion produced by omitting the adjective linear.

In practice it is the change of momentum that is observed and the impulse is then determined from the impulse momentum equation. If the time the force is acting is known then, assuming the force is constant, the force could be estimated from the impulse momentum equation.

Suppose that the speed of a car of mass 1000 kg moving with speed 20 ms$^{-1}$ suddenly drops, due to braking, to 15 ms$^{-1}$. The impulse momentum equation shows that an impulse of magnitude 5000 Ns has been applied but more information would be necessary to find the force producing this change. If the same braking action were carried out (i.e. the same impulse applied) at a speed of 18 ms$^{-1}$ then the momentum becomes

$$(18000 - 5000) \text{ Ns} = 13000 \text{ Ns}$$

so that the speed drops to 13 ms$^{-1}$.

It is very unlikely that the force will be constant throughout contact and therefore a generalisation of the definition of impulse is necessary. As in the case of work the generalisation can be seen from a graphical interpretation of impulse. Both the following diagrams show the behaviour of the component of force in a given direction with time $t$.

## Impulse and momentum

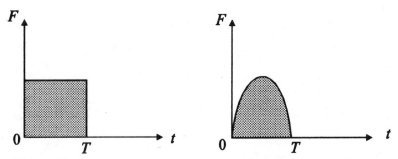

The left hand diagram shows the component to have a constant value whilst the right hand one shows variation with time; this is more likely to be a realistic form for a sharp blow since it vanishes at $t = 0$ and $t = T$. The impulse of the constant force is therefore the shaded area under the line and this suggests that the impulse for the variable force is also the area under the curve. This therefore suggests that the **impulse is defined as** $\int_0^T F \, dt$.

The next step is to see whether this can enable a change of velocity to be found simply in terms of the impulse.

Newton's law of motion gives
$$m \frac{dv}{dt} = F.$$

Integrating this equation from $t = 0$ and $t = T$ gives
$$(mv)_{t=T} - (mv)_{t=0} = \int_0^T F \, dt.$$

The left hand side of this equation is the change of momentum, the right hand side is the impulse acting and therefore the impulse momentum principle holds for all forces.

**Modelling a sharp blow**

In modelling a sharp blow it is not necessary to consider the precise form of the force but simply say that an "impulse" is applied so that there is an instantaneous change of momentum equal to the impulse. The model is therefore one where it is assumed that there are large forces acting for effectively infinitesimal periods of time so that the impulse, defined as the integral of force, is finite.

During any sharp blow there will obviously be forces like gravity acting but, in the limit of zero time, these do not make a contribution to the impulse. This is also true of the tension in an elastic spring or string, neither of which can sustain an impulsive tension. (The tension is proportional to the extension, the impulsive tension would therefore be the integral of the extension over a zero interval and is therefore zero.) The forces which make a contribution in the limit as time tends to zero are referred to as "impulsive" forces and normally their detailed behaviour is not known. There are no functions, in the normal sense of the word, whose integral over a zero interval is non-zero but it is possible to construct some bounded functions whose integral over a very small interval is finite and non-zero. One such function is $\dfrac{T}{t^2 + T^2}$. You will not yet have covered calculating its integral from $t = 0$ to $t = T$, but it can be shown to be equal to $\dfrac{\pi}{2}$. You might like to try and plot it in the range $0 \le t \le T$, for small values of $T$.

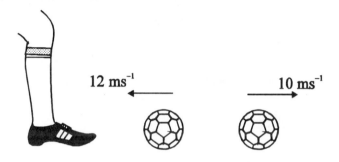

The diagram shows a football, of mass 0.4 kg, moving horizontally with speed 12 ms$^{-1}$ when it is kicked by a footballer so that immediately after leaving him it is moving horizontally with speed 10 ms$^{-1}$. The momentum of the ball after the kick is 4 Ns to the right and before the kick it is 4.8 Ns to the left. Therefore the change in the momentum to the right is 8.8 Ns and this is the impulse applied by the kick.

There is no way you could find the actual force without making more assumptions about its form (see Exercise 7.1, questions 10 and 11) but the important thing however is that, even without knowing the precise form of the force, you could now work out what happens when the ball arrives with a velocity of 8 ms$^{-1}$. The modelling assumption is that the player kicks exactly as before i.e. that the impulse of the force of the kick is 8.8 Ns. The football is assumed to move away with a speed of $v$ ms$^{-1}$, so that the momentum to the right afterwards is $0.4v$ N s and before it was $-3.2$ Ns. The impulse momentum principle gives

$$0.4v + 3.2 = 8.8,$$

so that
$$v = 13.75.$$

The impulse associated with a force can obviously be calculated for any force and in fact can be used to solve other problems of motion. It is not a particularly good method for solving other problems and it is really best to just try and keep the idea that impulse is something particularly linked to sharp blows and is proportional to the change of momentum. You should also try and remember that impulse as force × time is a very special case and one which is very unlikely to occur particularly in real problems involving sharp blows.

In solving problems involving impulse and momentum you should remember that, as for problems involving Newton's equations of motion, it is important to pick a reference direction.

## Example 7.1

A ball of mass 0.05 kg hits a vertical wall with a speed of 10 ms$^{-1}$, the direction of motion being perpendicular to the wall. The ball rebounds back with a speed of 8ms$^{-1}$. Find the impulse exerted by the wall on the ball.

The positive direction is taken to be that that away from the wall as shown in the diagram. The linear momentum of the ball immediately after impact is 0.05 × 8 Ns = 0.4 Ns, the linear momentum immediately before is − 0.05 × 10 Ns = − 0.5 Ns.

The impulse is therefore 0.4 Ns − (−0.5) Ns = 0.9 Ns.

The impulse is, as you would expect, acting away from the wall. By Newton's third law there will be an equal and opposite impulse acting on the wall.

## Example 7.2

A cricket ball of mass 0.5 kg moving horizontally with speed 6 ms$^{-1}$ is struck by a bat which applies a horizontal impulse of magnitude 5 Ns and directed in the opposite direction to that of the ball. Find the velocity of the ball immediately after impact.

## Impulse and momentum

The velocities before and after impact, and the impulse, are assumed to be as shown in the diagram. The momentum before impact is $-0.5 \times 6$ Ns and that after is $0.5v$ Ns. Applying the impulse momentum principle gives

$$0.5v + 0.5 \times 6 = 5,$$

so that $\qquad v = 4.$

### Example 7.3

A ball of mass 0.3 kg is moving with speed 5 ms$^{-1}$ just as it hits a horizontal floor and bounces off the floor with a speed of 2 ms$^{-1}$. Find the impulse exerted on the ball by the floor assuming
(i) that the time of contact with the floor may be neglected,
(ii) the time of contact to be 0.05 s.

The momentum of the ball immediately after impact is 0.6 Ns upwards and immediately before impact it is $-1.5$ Ns. The total change in the momentum upwards is 2.1 Ns and this is the impulse applied to the ball. If the time of contact may be neglected then the impulse due to gravity may be neglected and the impulse exerted by the floor is 2.1 Ns.

The total force acting is $(F - 0.3 \times 9.8)$ N, where the force exerted by the floor is $F$ N.

The total impulse is therefore $\int_0^{0.05} (F - 0.3 \times 9.8) \, dt$ Ns $= \int_0^{0.05} F \, dt$ Ns $- .15$ Ns.

The integral is the impulse exerted by the floor, the total impulse is 2.1 Ns and therefore the impulse exerted by the floor is 2.25 Ns.

### Exercises 7.1

**1** Find the momentum of the following
(a) a particle of mass 0.03 kg moving with velocity 4 ms$^{-1}$,
(b) a cricket ball of mass 0.5 kg moving with velocity 15 ms$^{-1}$,
(c) a car of mass 1200 kg moving with speed 25 ms$^{-1}$.

Questions 2 to 4 refer to a particle of mass $m$ kg whose velocity changes from $u$ ms$^{-1}$ to $v$ ms$^{-1}$. Find the change in momentum.

**2** $m = 0.5, u = 4, v = 6$.
**3** $m = 1.6, u = 4, v = -2$.
**4** $m = 2.2, u = -3, v = -5$.

## Impulse and momentum

**5** The kinetic energy of a particle of mass 0.4 kg is 3.2 J, find the magnitude of its momentum.

**6** An impulse of magnitude 3.2 Ns is applied to a particle of mass 0.8 kg at rest. The particle is free to move along the $x$-axis and the impulse is applied in this direction. Find the resulting speed of the particle.

**7** A railway truck of mass 1400 kg, moving along a straight horizontal track at 6 ms$^{-1}$, rebounds from fixed buffers with a speed of 1.5 ms$^{-1}$.
Find the impulse exerted by the buffers on the truck.

**8** An ice skater whose total mass is 70 kg receives a horizontal impulse of magnitude 200 Ns when standing at rest. Find the initial speed of the skater and, given that the total resistance is 25 N, determine the total distance moved by the skater.

**9** A tennis ball of mass 0.08 kg moving horizontally towards the racquet with speed 6 ms$^{-1}$ is hit by a racquet and leaves the racquet horizontally with a speed of 12 ms$^{-1}$. Calculate the magnitude of the impulse on the ball.

**10** Assuming that the ball and racquet in the previous exercise are in contact for 0.04 s find the force acting assuming
(i) that it is constant,
(ii) that it is of the form $ct$ N, where $t$ is time, in seconds, measured from impact and $c$ is a constant.

**11** A cricket ball of mass 0.15 kg moving horizontally with speed 14 ms$^{-1}$ just as it reaches a batsman is hit straight back horizontally with a speed of 24 ms$^{-1}$. Find the impulse of the bat on the ball.

**12** Assuming that the bat and the ball in the previous exercise are in contact for 0.05 s find the form of the force exerted by the bat on the ball assuming that it can be expressed as $kt(0.05 - t)$ N, where $k$ is a constant.

**13** A ball of mass 0.2 kg falls vertically onto a horizontal floor which it strikes with a speed of 10 ms$^{-1}$ and bounces to reach a height of 2.5 m above the floor.
Find the impulse exerted by the floor.

**14** A ball of mass 0.2 kg falls vertically and when moving with speed 6 ms$^{-1}$ is struck by a bat which is moving vertically upwards so that after the ball leaves the bat it has a speed of 4 ms$^{-1}$ vertically upwards. Find, assuming that the bat and ball are in contact for 0.05 s, the impulse exerted by the bat.

## 7.2 Conservation of momentum

When a batsman strikes a ball there will, by Newton's third law, be an opposite impulse acting on the bat, and therefore, on the batsman. This should affect his momentum but normally he would exert a further impulse on the ground to stay at rest. This kind of reaction is particularly obvious in the case of shooting a rifle when there is a recoil. There are however many problems involving sharp blows between bodies where one body does not compensate like a human ball player and motion of both bodies is possible. Obvious examples are snooker balls, or cars colliding. An example of such a problem is that of the collision of two balls moving directly towards each other as in the diagram.

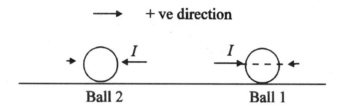

In the collision there will be impulsive forces acting between the balls during the collision. By Newton's third law the forces exerted by ball 2 on ball 1 during collision are equal and opposite to those exerted by ball 1 on ball 2. Since impulse is the integral of the force the impulses acting on the balls are equal and opposite. The impulse on ball 1 due to ball 2 is denoted by $I$ and that of ball 2 due to ball 1 is therefore $-I$ (i.e. $I$ to the left).

The positive direction is taken to be towards the right so that by the impulse momentum principle

Change in momentum of ball 1 = $I$,

Change in momentum of ball 2 = $-I$.

Adding these gives that the change in total momentum of balls 1 and 2 = 0.

Therefore the total momentum of the two balls is unchanged (i.e it is conserved), this is also true for any number of particles and this is the **principle of conservation of momentum** which states that

**During any period when there are no external impulses acting on a system of interacting particles the total momentum remains constant.**

It is also possible to prove for a system when external impulses are present a **general impulse momentum principle** which states that

**If impulses are applied to a system of interacting particles the change in momentum due to the impulse is equal to the sum of the impulses applied.**

Three different types of situations involving impulsive motion of a system of particles will be considered

(i) Collisions between bodies which move together after collision; these are called inelastic collisions and can usually be solved by use of the principle of conservation of linear momentum. These problems are considered in detail in 7.3.

(ii) Collisions where the bodies bounce apart after collision; these are called elastic collisions and to solve them it is necessary to know something about the elastic properties of the bodies. The method for solving these problems is given in 7.5.

(iii) Problems where the bodies are connected together by an inelastic string so that they move together. A simple example is two cars connected by a tow rope and one moving off. Apart from the simple problems of two bodies moving along a line some problems involving particles connected by a string passing over a pulley are also considered.

All these problems are examined in 7.4.

## 7.3 Inelastic collisions

Problems such as those when two cars collide with each other are usually quite easy to solve using the principle of conservation of momentum. The total momentum before collision will be known, you have to be very careful to pick a reference direction so that you get the correct signs for the momentum. The method is best seen by working through the following examples.

### Example 7.4

A car of mass 1400 kg moving with a speed of 4 ms$^{-1}$ crashes into a stationary car of mass 1000 kg. After collision they move together. Find the common speed immediately after collision.

The reference direction is taken in the direction of the moving car as shown in the diagram and the common speed immediately after collision is denoted by $v$ ms$^{-1}$ as shown. The momentum before collision is therefore $1400 \times 4$ Ns = 5600 Ns. The momentum after is $(1400 + 1000)v$ Ns $= 2400v$ Ns. Conservation of momentum gives
$$2400v = 5600,$$
so the cars move together with speed $\frac{7}{3}$ ms$^{-1}$.

## Example 7.5

Two cars of mass 1200 kg and 1300 kg are moving directly towards each other with speeds of 12 ms$^{-1}$ and 16ms$^{-1}$ respectively. After collision the two cars move together. Find their common speed.

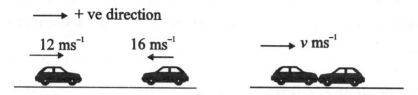

The reference direction is taken to be that of the initial motion of the lighter car so that the momentum of the latter is 14400 Ns whilst that of the heavier car is $-$ 20800 Ns, the total momentum before collision is therefore $-$ 6400 Ns.

It is now assumed that both move together with speed $v$ ms$^{-1}$ to the right. The total linear momentum after collision is $2500v$ Ns and this is equal to $-$ 6400 Ns so $v = -$ 2.56 and therefore both cars move to the left with speed 2.56 ms$^{-1}$.

## Exercises 7.2

Questions 1 to 6 refer to a particle of mass $M$ kg moving with velocity $u$ ms$^{-1}$ colliding with a particle of mass $m$ kg moving with velocity $v$ ms$^{-1}$. After collision they move together with velocity $w$ ms$^{-1}$.

**1** $M = 0.4$, $m = 0.2$, $u = 3$, $v = 0$, find $w$.
**2** $M = 0.2$, $m = 0.6$, $u = 5$, $v = 0$, find $w$.
**3** $M = 4$, $m = 3$, $u = 5$, $v = 2$, find $w$.
**4** $M = 0.8$, $m = 0.6$, $u = 2$, $v = 5$, find $w$ and the total change in the kinetic energy.
**5** $M = 3$, $m = 2$, $u = 5$, $v = -2$, find $w$ and the total change in the kinetic energy.
**6** $M = 2$, $m = 3$, $u = -4$, $v = 7$, find $w$ and the total change in the kinetic energy.

**7** Car B, which is of mass 1200 kg, is initially stationary and is struck by car A, of mass 1500 kg, moving with speed 15 ms$^{-1}$. The cars become entangled and move together immediately after collision. Find their common speed after collision.

**8** A railway truck of mass 12 tonnes moving with speed 2 ms$^{-1}$ collides with a stationary truck mass 16 tonnes. The trucks move together immediately on impact. Find their common speed.

**9** A railway goods waggon $A$ of total mass 20 tonnes is moving along the horizontal track in a railway yard at a speed of 1.5 kmh$^{-1}$. A second goods waggon $B$ with a total mass of 25 tonnes and moving with speed 3 kmh$^{-1}$ overtakes it and is coupled to it. Find the common velocity $v$ of the two waggons as they move together after being coupled.

**10** A 0.045 kg rifle bullet is fired horizontally with a velocity of 425 ms$^{-1}$ into a 5 kg block of wood which can move freely in the horizontal direction. Determine the final velocity of the block.

**11** A girl of mass 44 kg runs and, when her horizontal velocity is 5 ms$^{-1}$, jumps on a stationary 16 kg sledge. The girl and sledge travel a distance of 18 m horizontally on snow before coming to rest. What is the coefficient of sliding friction between the toboggan and the snow?

## 7.4 Impulsive motion of connected particles

**Particles moving along the same line**

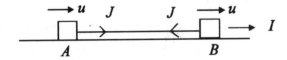

The typical problem is as shown in the diagram, two particles $A$ and $B$, of mass $m$ and $M$, respectively are connected together by a light inextensible string and, when the string is taut, an impulse $I$ is applied to $B$ in the direction from $A$ to $B$. Both particles will move with the same speed $u$ as shown and therefore there will be an impulse $J$ acting at $A$ in the direction $A$ to $B$. By Newton's third law there will be an impulse of equal magnitude acting at $B$ from $B$ to $A$. This impulse which is acting along the string is usually called the impulsive tension even though it is not a force.

The total applied impulse is $I$ and this by the impulse momentum principle for a system of particles is equal to the total change in momentum i.e. $(m + M)u$. Therefore

$$u = \frac{I}{m + M}.$$

The impulsive tension can now be found by applying the impulse momentum principle to particle $A$, i.e. $mu = J$, so that 
$$J = \frac{mI}{m + M}.$$

An alternative method would have been to apply the impulse momentum to both $A$ and $B$. The equation obtained for $B$ would be $I - J = Mu$. Eliminating $J$ gives $u$ as before. It is simpler for problems of this type, as for problems involving the motion of two particles, to consider the whole system first and then one of the particles.

Some care has to be used in considering the impulsive motion of connected particles.

The diagram shows the two particles $A$ and $B$ discussed above moving with speed $u$ in the direction $AB$ when an impulse of magnitude $I$ is applied to $B$ in the sense from $A$ to $B$.

## Impulse and momentum

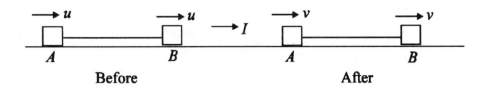

Before            After

After the impulse has been applied they will both be moving with a new speed $v$ and the impulse momentum principle for a system of particles gives
$$(m + M)(v - u) = I.$$
Therefore $v$ can be found.

If the impulse had been applied from $B$ to $A$ and it had been assumed that both particles still moved together then this would have given that they would both be moving to the left with speed $v$ where
$$(m + M)(v + u) = I.$$
This would mean that there would be an impulsive tension to the left on the particle $A$. This is impossible and therefore in this case the impulse would only affect the particle $B$.

### Example 7.6

Car $A$ in the diagram is about to tow car $B$. The tow rope is slightly slack so that car $A$ can reach a speed of 2.5 ms$^{-1}$ before the rope tightens. Determine the motion of the cars immediately after the rope tightens. The masses of cars $A$ and $B$ are 1500 kg and 1250 kg respectively.

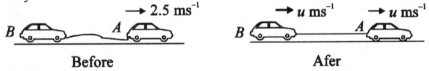

Before            Afer

Immediately after the rope tightens the cars will be moving at the same speed of $u$ ms$^{-1}$. In this case there is no applied impulse and the principle of conservation of momentum may be applied

The principle of conservation of momentum gives
$$1500 \times 2.5 = (1500 + 1250)u$$
so that $u = 1.36$. The impulsive tension tension in the rope is $1250u$ Ns = 1700 Ns.

### Example 7.7

Two particles $A$ and $B$ of mass $2m$ and $6m$ lie in a straight line, joined by an inextensible string, which is just taut. An impulse of magnitude $16mU$ is applied to $B$ in the sense from $A$ to $B$. Determine the subsequent motion.

# Impulse and momentum

The only possible motion will be along the string and since the string is inextensible both particles move with the same speed $v$. The total momentum after the particles start moving is $2mv + 6mv = 8mv$. The impulse momentum principle gives
$$8mv = 16mU,$$
so that $\qquad v = 2U.$

The impulsive tension acts on $A$ towards $B$ and is denoted by $I$.

Applying the impulse momentum principle to $A$ gives $I = 2mv = 4mU$.

If the original impulse had been from $B$ to $A$ then there would have been no impulsive tension in the string, so $A$ would not move and $6mv = 16mU$, giving
$$v = \frac{8U}{3}.$$

## Problems involving pulleys

Problems involving particles on a string passing over a pulley can be solved by considering the system as a whole, as for the case of general motion involving pulleys, but it is not really a safe approach. It is better to apply the impulse momentum principle carefully to each particle. The most commonly occurring types of problem are illustrated in the following examples.

## Example 7.8

Two particles of mass 0.4 kg and 0.6 kg attached one at each end of an inextensible string passing over a smooth pulley, as shown in the diagram, are set in motion. At the instant when the particles have a common speed of 3 ms$^{-1}$ the heavier particle hits a horizontal surface off which it does not rebound. (The surface is referred to as being inelastic.) Find the speed with which the heavier particle is first jerked into motion.

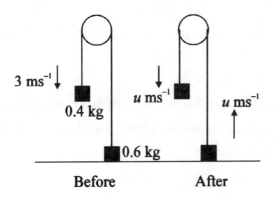

Once the heavier particle has stopped the lighter one continues upwards under gravity to its highest point and then falls reaching the point where the string is about to become taut with speed 3 ms$^{-1}$. When the string becomes taut there will be an impulsive tension $T$ Ns in the string and the heavier particle will jerked off the plane and both particles will start to move with a common speed $u$ ms$^{-1}$.

Applying the impulse momentum principle to the heavier particle gives $T = 0.6u$.

The impulsive tension acting on the lighter particle will be upwards, the velocity of the particle will have changed from 3 ms$^{-1}$ downwards to $u$ ms$^{-1}$ downwards. Therefore the change in linear momentum downwards is $0.4(u - 3)$ and this is equal to $-T$. Eliminating $T$ gives
$$0.4(u - 3) + 0.6u = 0$$
i.e
$$0.4u + 0.6u = 1.2$$
so that the heavier particle is jerked off with speed 1.2 ms$^{-1}$.

The equation for $u$ is exactly the same as if the two particles had been in a straight line and the principle of conservation of momentum applied. That method should not be used to solve a problem in an examination as it needs careful justification but it could help as a check.

**Example 7.9**

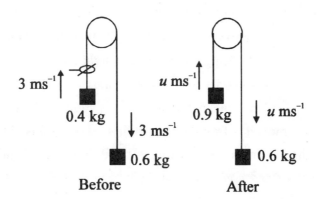

Before            After

When the particles are moving freely with a speed of 3 ms$^{-1}$ as shown in the diagram the lighter particle picks up a mass 0.5 kg which is lying on a fixed ring through which the particle passes. Find the common speed of the system immediately after the mass has been picked up.

The particles move with a new unknown speed $u$ ms$^{-1}$.

There will be an impulsive tension $T$ in the string and this is the impulse acting on the combined particle so, applying the impulse momentum principle to this combined particle, $T = (0.5 + 0.4)u - 3 \times 0.4$. The velocity of the heavier particle, downwards, will change from 3 ms$^{-1}$ to $u$ ms$^{-1}$ so the change in its linear momentum downwards is $0.6u - 0.6 \times 3$ and this is equal to $-T$. Eliminating $T$ gives

$$(0.5 + 0.4)u - 0.4 \times 3 + 0.6u - 0.6 \times 3 = 0$$

i.e. $$(0.5 + 0.4)u + 0.6u = 0.4 \times 3 + 0.6 \times 3,$$

and $$u = 2.$$

## Exercises 7.3

**1** Two particles $P$ and $Q$ of mass 4 kg and 3 kg respectively lie on a smooth table connected together by a light inextensible string. Particle $P$ is projected away from $Q$ with speed 8 ms$^{-1}$. Find the common speed of the particles after the string becomes taut and the impulsive tension in the string.

**2** Particle $A$, of mass 0.2 kg, lies at rest on a smooth horizontal table and at a distance of 0.4 m from its edge. The surface of the table is at a height of 2 m above the floor. Particle $A$ is joined by a light inelastic string of length 0.9 m to a second particle $B$ of mass 0.4 kg. This particle is placed at the edge of the table and then pushed over the edge in such a way that the string is perpendicular to the edge of the table. Find the speed of $A$ when it starts moving and also the impulsive tension in the string.

Questions 3 to 6 refer to two particles $A$ and $B$, of masses $m$ kg and $M$ kg respectively, connected by a light inextensible string passing over a light smooth pulley.

**3** $m = 0.3$, $M = 0.2$, the system is set off from rest and, when both particles are moving with speed 2 ms$^{-1}$, the particle $B$ picks up from rest an additional particle of mass 0.3 kg. Find the further distance moved before the system first comes to instantaneous rest.

**4** $m = 0.5$, $M = 0.3$, the system is set off from rest and, after descending 5 m, the particle $A$ strikes an inelastic floor and comes to rest. Find the time that it remains on the floor and the speed with which it is jerked off.

**5** $m = 0.8$, $M = 0.4$, when both masses are moving with speed 2 ms$^{-1}$ $A$ passes through a small ring and a mass of 0.5 kg is removed from it. Find the time that elapses before $A$ next passes through the ring.

**6** $m = 0.5$, $M = 0.4$, the system is at rest with $A$ resting on a smooth plane. A falling particle of mass 0.2 kg moving with speed 4 ms$^{-1}$ strikes $B$ and sticks to it. Find the height to which $A$ rises.

## 7.5 Elastic collisions

There are effectively two different situations, one where one body remains fixed (ball hitting a wall) and the other where both bodies can move. They will be examined in turn.

**One body fixed**

When a ball, for example, hits a wall there is no theoretical method for finding the speed just after leaving the wall. Experiments have however shown that if a ball, or particle, is moving with speed $u$ perpendicular to the wall just before impact then the speed with which it leaves the wall is $eu$, where $e$ is a number depending on the elastic properties of both the wall and the ball and is known as the coefficient of restitution. This experimental law was first established by Newton.

The speed $u$ with which the ball is approaching the wall is referred to by Newton as the speed of approach, and he referred to $v$, the speed of the ball as it leaves the wall, as the speed of separation. He expressed his experimental law in the form

$$\frac{\text{speed of separation}}{\text{speed of approach}} = e,$$

where $e$ is the coefficient of restitution and satisfies the conditions $0 \leq e \leq 1$. The lower limit refers to a perfectly plastic, or inelastic collision, where the ball sticks to the wall. The upper limit corresponds to a perfectly elastic collision where the ball leaves the wall with speed $u$.

If the speed before collision is $u$ then the speed after collision is $eu$ and for a ball of mass $m$ there is an energy loss of $\frac{1}{2}mu^2 (1-e^2)$. There is therefore an energy loss in any collision for $e < 1$. Some of the energy loss is dissipated as heat (e.g. a squash ball gets very warm).

**Example 7.10**

A ball is dropped vertically downwards onto a smooth plane from a height of 1.4 m. Given that the coefficient of restitution between the ball and plane is 0.6 find
(i) the height to which the ball first bounces,
(ii) the time taken from dropping the ball to it reaching the top of its first bounce.

The formula $v^2 = u^2 + 2gh$, gives that the speed, in ms$^{-1}$, of the ball when it first reaches the plane is $\sqrt{2 \times 9.8 \times 1.4} = 5.24$.

The speed of rebound is, from Newton's law,
$$0.6 \times 5.24 = 3.14.$$

Applying the above formula again gives the height, in m, to which the ball rises as
$$\frac{3.14^2}{2 \times 9.8} = 0.50.$$
Applying the formula $v = u + gt$ gives the time, in seconds, of the downwards motion as $\frac{5.24}{9.8} = 0.53$. Applying the formula to the upwards motion gives the time, in seconds, to the top of the bounce as $\frac{3.14}{9.8} = 0.32$. The total time is therefore 0.85 s.

**Both bodies free to move**

Newton carried out experiments involving moving objects and found that his experimental law in the form
$$\frac{\text{speed of separation}}{\text{speed of approach}} = e, \quad \text{was still valid.}$$
Some care is needed in interpreting this when both bodies are moving.

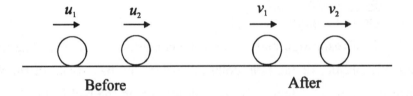

Before          After

The diagram shows two bodies moving along a line with velocity components $u_1$ and $u_2$. For $u_1 > u_2$ they will collide and the corresponding components after collision are $v_1$ and $v_2$. The speed of approach is the rate at which the distance between the two is decreasing before collision and this is $u_1 - u_2$. The speed of separation is the rate at which the distance between the two is increasing after collision and this is $v_2 - v_1$. Then Newton's law can be written as
$$v_2 - v_1 = e(u_1 - u_2) = -e(u_2 - u_1).$$
In solving practical problems it is important to pick a reference direction and calculate all the components in that direction. It is unwise to try and guess whether after collision a body is moving in a particular direction and show this direction in a diagram. If it turns out that a body is moving in the opposite direction to the reference one then this will show in calculations by the component being negative. It is easier to remember Newton's law in the form
$$v_2 - v_1 = -e(u_2 - u_1),$$
since always putting the minus sign outside avoids you having to remember to change the order of subtraction on the two sides of the equation.

## Impulse and momentum

In a collision problem the total momentum is conserved. This means that, if the masses of the two particles above are denoted by $m_1$ and $m_2$, then

$$m_1 u_1 + m_2 u_2 = m_1 v_1 + m_2 v_2.$$

All problems involving elastic collisions reduce to solving two equations of the above type.

### Example 7.11

A small smooth sphere of mass 0.1 kg moving with speed of 20 ms$^{-1}$ on a horizontal plane catches up and collides with a smooth sphere of the same radius but of mass 0.9 kg and moving with a speed of 5 ms$^{-1}$. The coefficient of restitution is $\frac{1}{3}$. Find the speeds of the spheres immediately after collision.

The most important step is to draw diagrams to show the motion before and after collision and to mark in a reference direction. There is no real point in trying to guess the directions of motion after collision and it is best to take all unknowns in the reference direction since it avoids problems with signs.

Before: 0.1 kg moving at 20 ms$^{-1}$, 0.9 kg moving at 5 ms$^{-1}$
After: $v_1$ ms$^{-1}$, $v_2$ ms$^{-1}$

In the diagram the velocities of the spheres after collision are $v_1$ ms$^{-1}$ and $v_2$ ms$^{-1}$ to the right.

The component of the momentum, in Ns, to the right before collision is

$$0.9 \times 5 + 0.1 \times 20 = 6.5 \quad \text{and after collision it is} \quad 0.1 v_1 + 0.9 v_2.$$

Conservation of momentum gives

$$6.5 = 0.1 v_1 + 0.9 v_2.$$

Newton's law gives $\quad v_2 - v_1 = -\frac{1}{3}(5 - 20) = 5.$

Solving these gives $\quad v_1 = 2, \quad v_2 = 7.$

The kinetic energy change can also be worked out for this example.

The kinetic energy before collision is $\frac{1}{2}(0.1 \times 20^2 + 0.9 \times 5^2)$ J = 31.25 J.

The kinetic energy after collision is $\frac{1}{2}(0.1 \times 2^2 + 0.9 \times 7^2)$ J = 22.25 J.

There is therefore a loss of kinetic energy in this collision of 9 J.

## Example 7.12

Two small smooth spheres of equal radius but of mass 0.3 kg and 0.1 kg are moving directly towards each other with speeds of 4 ms$^{-1}$ and 6 ms$^{-1}$ respectively, the coefficient of restitution being 0.5. Find their speeds after collision.

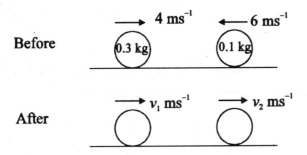

The diagram shows the velocities before and after collision. In calculating the momentum and using Newton's law it is important to remember that the component of the initial velocity of the lighter sphere is $-6$ ms$^{-1}$ to the right. The component of the momentum, in Ns, to the right before collision is $0.3 \times 4 - 0.1 \times 6 = 0.6$ and after collision it is $0.3v_1 + 0.1v_2$.

Conservation of momentum gives
$$0.3v_1 + 0.1v_2 = 0.6.$$

Newton's law gives
$$v_2 - v_1 = -0.5(-6-4) = 5,$$
Solving these equations gives $v_1 = 0.25$, $v_2 = 5.25$.


## Exercises 7.3

Questions 1 to 4 refer to a small ball dropped from rest at a height $h$ onto a smooth plane, the coefficient of restitution being $e$, and $h_1$ being the height reached after the first bounce.

**1** $h = 5$ m, $e = 0.4$, find $h_1$.

**2** $h = 8$ m, $h_1 = 5$ m, find $e$.

**3** $h = 1$ m, $e = \frac{1}{4}$ find the total distance travelled by the ball before it comes to rest.

**4** $h = 2.4$ m, $e = 0.4$; find, given that the ball is of mass 0.2 kg, the magnitude of the impulse on the ball at the first bounce.

**5** A tennis ball is projected vertically downwards from a height of 1.6 m onto a tennis court. Given that the coefficient of restitution is 0.8 find the speed of projection in order that the ball just returns to the point of projection after bouncing once.

In questions 6 to 8 a smooth sphere of mass $m_1$ kg moving with speed $u_1$ ms$^{-1}$ overtakes and collides directly with a smooth sphere of mass $m_2$ kg moving with speed $u_2$ ms$^{-1}$ in the same direction. The velocity components after impact and measured in the original direction of motion are denoted by $v_1$ ms$^{-1}$ and $v_2$ ms$^{-1}$, respectively.

**6** $m_1 = 3$, $m_2 = 1$, $u_1 = 6$, $u_2 = 1$, $e = 0.4$, find $v_1$ and $v_2$.

**7** $m_1 = 2$, $m_2 = 3$, $u_1 = 6$, $u_2 = 2$, $v_1 = 3$, find $e$ and $v_2$.

**8** $m_2 = 10$, $u_1 = 9$, $u_2 = 2$, $v_1 = 2$, $v_2 = 5$, find $e$ and $m_1$.

In questions 9 to 11 a smooth sphere of mass $m_1$ kg moving with speed $u_1$ ms$^{-1}$ collides directly with a smooth sphere of mass $m_2$ kg moving with speed $u_2$ ms$^{-1}$ in the opposite direction. The velocity components after impact and measured in the direction of motion of the first sphere are denoted by $v_1$ ms$^{-1}$ and $v_2$ ms$^{-1}$, respectively.

**9** $m_1 = 4$, $m_2 = 1$, $u_1 = 3$, $u_2 = 1$, $e = 0.5$, find $v_1$ and $v_2$.

**10** $m_1 = 4$, $u_1 = 3$, $u_2 = 5$, $v_1 = 2$, $e = 0.2$, find $m_2$ and $v_2$.

**11** $m_2 = 12$, $u_1 = 10$, $u_2 = 2$, $v_1 = 2$, $v_2 = 4$, find $e$ and $m_1$.

**12** A smooth sphere of mass 4 kg collides directly with a smooth sphere of mass 8 kg which is at rest. Find the condition to be satisfied by $e$ in order that the spheres move in opposite directions after collision.

**13** Identical cars $A$, $B$, $C$ are in a straight line, very slightly apart, with their brakes off. Car $A$ is pushed towards the others so that it hits car $B$ with speed 1 ms$^{-1}$. Given that the coefficient of restitution is 0.75 find the speeds of the cars after all collisions have finished.

## **Miscellaneous Exercises 7**

**1** The police are investigating an accident where a moving car of mass 1000 kg collided directly with a parked one of mass 1200 kg. Immediately after the collision the cars became enmeshed and slid forward together a distance of 10 m. Assuming that the coefficient of friction is 0.5, that the road is level and that the only horizontal force acting on the enmeshed cars is friction find the common speed of the cars as they start moving. Hence find the speed of the moving car at the instant of collision.

**2** A lorry of mass 4 tonnes is towing a car of mass 1200 kg. The lorry sets off with the tow rope slack but when its speed is 1.5 ms$^{-1}$ the tow rope tightens. Find
(i) the speed of the car immediately it starts moving,
(ii) the impulsive tension in the tow rope.

**3** A car of mass 1200 kg moving with speed 30 ms$^{-1}$ crashes into the back of a car of mass 1000 kg moving in the same direction with speed 15 ms$^{-1}$. After collision the cars move together. Find their common speed immediately after the collision and the impulse on the lighter car.

4

The diagram shows two cars $A$ and $B$ of masses 1200 kg and 1800 kg on a horizontal road. Car $A$ has broken down and car $B$ is about to tow it. The speed of car $B$ just as the tow rope tightens is 5 ms$^{-1}$. Modelling the cars as particles and assuming that the tow rope (whose weight may be neglected) is parallel to the road and along the line joining the cars find

(i) the common speed of the two cars immediately after car $A$ has started moving,

(ii) the impulsive tension in the tow rope.

The two cars then continue to move along the road (which is horizontal) until they reach a steady speed of 15 ms$^{-1}$. They then continue at this constant speed. Given that car $B$ is working at a constant rate of 20 kW find the total resistance acting on the two cars.

Given further that the resistance on each vehicle is proportional to its mass find the tension in the tow rope.

5 Two particles $P$ and $Q$ of masses $4m$ and $5m$ respectively are attached one to each end of a light inextensible string whiuch passes over a small smooth pulley. The particles move in a vertical plane with both hanging parts vertical and they are released from rest. Find in terms of $m$ and/or $g$ as appropriate the magnitudes of the accelerations of the particles and the tension in the string.

When the particle $P$ is moving upwards with speed $V$ it picks up from a point $A$ an additional particle of mass $2m$ so as to form a composite particle $R$ of mass $6m$. Find the initial speed of $R$.

6

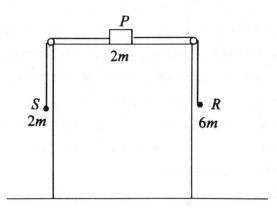

The diagram shows a particle $P$ of mass $2m$ on a rough horizontal table and attached by light inextensible strings to particles $R$ and $S$ of mass $6m$ and $2m$ respectively. The coefficient of friction between $P$ and the table is 0.5. The strings pass over light smooth pulleys on opposite sides of the table so that $R$ and $S$ can move freely with the strings perpendicular to the table edges. Given that the system is released from rest find the magnitude of the common acceleration of the particles and the tension in the string joining $P$ and $S$.

After falling a distance $d$ from rest, the particle $R$ strikes an inelastic floor and is brought to rest. Find, for the period after $R$ strikes the floor, the further distance that $S$ rises.

Find also, assuming that in the subsequent motion $P$ remains on the table and $S$ never reaches the table, the speed at which $R$ is jerked off the floor.

**7** A sphere $P$ of mass $m$ moving with constant speed $5u$ catches up and collides directly with an identical sphere $Q$ moving with speed $2u$. The kinetic energy lost in this collision is $2mu^2$. Find the speed of $P$ immediately after colliding with $Q$.

**8** A small smooth sphere is dropped from a point at a height of 0.6 m above a smooth horizontal floor. The sphere falls vertically, strikes the floor and bounces to a height of 0.15 m above the floor. Find

(a) the speed of the sphere when it hits the floor,

(b) the coefficient of restitution between the sphere and the floor.

**9** A particle of mass $m$ is dropped from rest from a point $A$ at a height $h$ above a horizontal surface. After hitting the surface the particle rebounds vertically to a height of $\frac{h}{2}$ to the point $B$. Find the loss of kinetic energy due to impact and the impulse on the particle. At the instant that the first particle hits the surface a second identical particle is dropped vertically from $A$. Show that the particles collide at $B$. The two particle coalesce after collision. Find the speed of the combined particle immediately after collision.

**10** A small smooth ball falls vertically on to a smooth horizontal floor. Its kinetic energy is reduced by one half by the impact. Find the coefficient of restitution.

**11** A small smooth sphere $P$ of mass $6m$ moving on a smooth plane in a straight line with constant speed $8u$ collides directly with a small smooth sphere $Q$ of the same radius but of mass $4m$ and moving in the same direction with speed $6u$. The direction of motion of sphere $Q$ is unchanged by the collision and immediately after the collision it moves with speed $8u$. Find

(i) the speed of $P$ immediately after collision,

(ii) the coefficient of restitution.

**12** Two small smooth spheres $A$ and $B$ of of equal mass moving in the same straight line (and in the same direction) collide directly. After the collision their directions of motion remain unchanged but their speeds are $v$ and $1.5v$ respectively. Show that the coefficient of restitution is 0.6.

**13** Sphere $P$ moves on a horizontal floor and collides directly with an identical sphere $Q$ which is at rest at a distance of 2 m from a smooth vertical wall, $Q$ being nearer to the wall than $P$. The motions before and after all possible collisions are perpendicular to the wall and the coefficient of restitution for all collisions is 0.6.

(i) Show that when $Q$ collides with the wall for the first time, $P$ is at a distance of 1.5 m fom the wall.

(ii) Find the distance of the spheres from the wall when they collide for the second time.

**14** Two small smooth spheres $P$ and $Q$ of equal radius but of mass $2m$ and $4m$ respectively are moving directly towards each other on a smooth horizontal table with speeds $2u$ and $3u$, respectively. The collision is such that $Q$ receives an impulse of magnitude $8amu$, where $a$ is a constant. Find

(a) the speeds of the spheres immediately after collision,

(b) the coefficient of restitution,

(c) the range of possible values of $a$,

(d) the value of $a$ such that the kinetic energy after collision is $6mu^2$.

**15** A small smooth sphere $A$ of mass $m$ and moving with speed $5u$ catches up and collides directly with a second sphere $B$ of mass $4m$ and moving with speed $u$. After collision the direction of motion of $A$ is reversed and its speed is $u$. Find

(a) the speed of $B$ after collision,

(b) the coefficient of restitution,

(c) the kinetic energy lost in the collision.

**16**

The diagram shows two small smooth spheres $A$ and $B$ of equal radii at rest on a horizontal table, with sphere $A$ between a smooth vertical wall and sphere $B$. The line joining the centres of the spheres is perpendicular to the wall. Sphere $A$ is of mass $m$ and sphere $B$ is of mass $qm$. Sphere $A$ is projected so as to collide directly with sphere $B$. Given that the coefficient of restitution between the spheres is $\frac{2}{5}$ and $q > \frac{5}{2}$, show that the direction of motion of sphere $A$ is reversed by the impact.

Given, additionally, that the coefficient of restitution between sphere $A$ and the wall is $\frac{1}{5}$ find the condition to be satisfied by $q$ so that the spheres will collide again.

(You may assume that, between collisions, the spheres move with constant speed.)

**17** Two small smooth spheres $P$ and $Q$ of equal radii but of masses $m$ and $3m$ respectively are moving towards each other on a smooth horizontal table. Before collision the speeds of $P$ and $Q$ are $3u$ and $6u$ respectively and after collision the direction of motion of $P$ is reversed and it moves with speed $5u$. Find the impulse on $Q$, the speed of $Q$ after collision and the coefficient of restitution between $P$ and $Q$.

**18** Two small smooth spheres $A$ and $B$ of mass $2m$ and $5m$ respectively and moving along $Ox$ collide. The velocity of $A$ immediately before collision is $5u$ in the positive $x$-direction and immediately after collision the velocities of $A$ and $B$ in the positive $x$-direction are $2u$ and $4u$ respectively. Determine

(i) the velocity of B immediately before collision,

(ii) the magnitude of the impulse on $B$,

(ii) the value of the coefficient of restitution.

**19** Two small smooth spheres $A$ and $B$ of equal radius, but of masses $m$ and $km$ respectively, are at rest on a smooth horizontal floor. The sphere $B$ lies between sphere $A$ and a smooth vertical wall and is at a perpendicular distance of 3 m from the wall. The line $AB$ is perpendicular to the wall. The coefficient of restitution for collisions between the spheres and between sphere $B$ and the wall is 0.2. Sphere $A$ is then projected directly towards sphere $B$ with speed $u$ ms$^{-1}$. Find the velocities of the spheres immediately after impact.

(i) Given that $k = 3$ find

    (a) the distance between $A$ and $B$ when $B$ hits the wall,

    (b) the distance from the wall to the point at which the spheres next collide.

(ii) Given that $k \neq 3$ find the least value of $k$ so that the spheres do not collide after $B$ hits the wall.

**20** Two small smooth balls $A$ and $B$ of of masses 0.2 kg and 0.1 kg respectively, are moving in a straight vertical line through a fixed point $O$. The balls collide at a point $P$ above $O$. Immediately before collision ball $A$ is moving downwards and ball $B$ is moving upwards. The coefficient of restitution between the balls is $\frac{3}{5}$ and immediately after collision ball $A$ moves upwards with a speed of 1 ms$^{-1}$ and ball $B$ moves downwards with a speed of 17 ms$^{-1}$. Find

(i) the speeds of both balls immediately before collision,

(ii) the impulse imparted to $A$ by the collision.

Explain why your results are consistent with both balls being projected from a point below $P$ with the same initial speed, with $A$ being projected before $B$.

# Chapter 8

# Motion under gravity in two dimensions

After working through this chapter you should
- be able, for motion in a plane, to find the magnitude and direction of velocity of a particle in two dimensions, given its components, and vice-versa,
- have a clear idea of the form of the path of a particle moving under gravity and be able to solve problems of a particle being projected from a point.

## 8.1 Basic kinematics

All the dynamical problems that you have come across so far have involved motion in a line where it is only possible to move backwards and forwards. The situation is not quite as simple for motion in a plane, for example a ball moving on a horizontal plane can move in an infinite number of different directions. It is therefore necessary to generalise the idea of velocity to problems involving two dimensional motion.

Velocity is defined to be something which completely represents the rate of change of position of a body, both the rate at which distance is covered and the direction in which the body is moving. It is not particularly easy at this stage, except for motion in a straight line, to give a clear definition of rate of covering distance though if you walk along any kind of curved path you are still aware of some kind of 'speed'. The basic point however is that velocity is something which has associated with it both a magnitude and a direction: its 'speed' and the direction of motion.

The velocity of a man moving with constant speed 2 ms$^{-1}$ at an angle θ North of East as in the left hand diagram is defined to be of magnitude 2 ms$^{-1}$ at an angle θ North of East.

## Motion under gravity in two dimensions

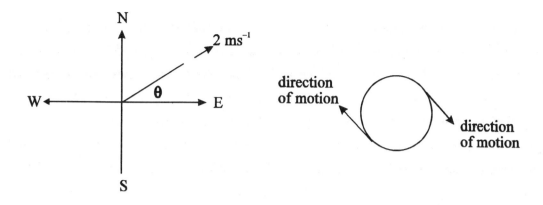

If the man were to walk round a circle as in the right hand diagram so that each small arc of the circle is described in equal time then the direction of motion at a point would be along the tangent to the circle. His velocity at any point on the circle is then defined to be along the tangent to the circle at that point and its magnitude to be the circumference of the circle divided by the time to describe a complete circle. A more precise definition of this magnitude will be given later.

Therefore the velocity at a point can be represented by a line, the direction of the line representing the direction of motion and its length representing the magnitude. Velocity is therefore, like force, a vector. This means that any velocity can be regarded as a combination of two, or more, separate motions or components. A simple example is the motion of rain when it is windy. If there is no wind the rain will fall vertically downwards but when it is windy it falls at an angle. This is because the actual velocity of the rain is a combination of the velocity due to falling under gravity and the wind velocity.

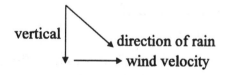

Velocities combine, like forces, according to the parallelogram rule.

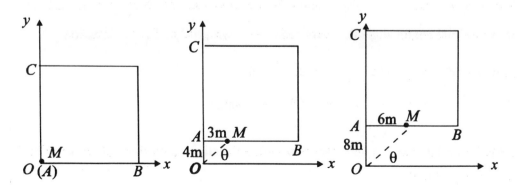

## Motion under gravity in two dimensions

You can see this by imagining a large board on an ice rink and with two perpendicular edges $AB$ and $AC$ pointing East and North. Initially the board is placed, as shown in the diagram, with $A$ at a fixed point $O$ and $AB$ and $AC$ parallel to fixed lines drawn East and North through $O$. These lines are taken to be coordinate axes $Ox$ and $Oy$. The board is then pulled northwards with speed 4 ms$^{-1}$ and at the same time a small animal $M$ is imagined to move along the line $AB$ with speed 3 ms$^{-1}$. After 1 s, $OA$ will be 4 m and $AM$ will be 3 m so that, referred to the axes $Ox$, $Oy$, $M$ will have coordinates (3,4) and $OM$ will be $\sqrt{4^2 + 3^2}$ m = 5 m. After 2 s the coordinates will be (6, 8) and $OM = 10$ m. Therefore $M$ will move along the line through $O$ and the point (3, 4), its speed will be 5 ms$^{-1}$. Therefore the motion of $M$ with speed 5 ms$^{-1}$ at an angle $\theta$ to $Ox$, where $\tan\theta = \frac{4}{3}$ is a combination of the two separate motions i.e. the components of the velocity are 3 ms$^{-1}$ along $Ox$ and 4 ms$^{-1}$ along $Oy$. This confirms the validity of the parallelogram of velocities in a simple case.

It is easier in more general motions to define the components in two perpendicular directions first rather than try and define directly something which represents the rate of change of position. If the components of velocity along $Ox$ and $Oy$ are defined by $u$ and $v$

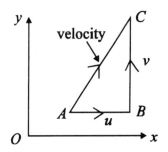

respectively then the velocity is represented, as shown in the diagram, by the hypotenuse of the right angled triangle $ABC$ where $AB$ is parallel to $Ox$ and proportional to $u$ and $BC$ is parallel to $Oy$ and proportional to $v$. The length of the line $AC$ represents the speed $w$ which is defined by $w = \sqrt{u^2 + v^2}$. The velocity is therefore defined to be of magnitude $w$ and in the direction making an angle $\theta$, where $\tan\theta = \frac{v}{u}$, with the positive $x$ direction.

From the right angled triangle $ABC$, $\frac{u}{w} = \cos\theta$, $\frac{v}{w} = \sin\theta$,

i.e. $\qquad\qquad\qquad u = w\cos\theta,\ v = w\sin\theta.$

The above equations let you convert the magnitude-direction definition of velocity to component form and vice-versa.

The equations are precisely the same as those relating the magnitude and direction of a force to its components and you have to be careful, as in Statics in 2.3, that you choose the correct quadrant for the direction of the velocity.

The position of a particle moving in a plane is specified completely by its $x$ and $y$ coordinates; these may depend on time. These coordinates are generally referred to as the displacements of the particle from the origin. The $x$- and $y$-components of velocity, $u$ and $v$ respectively, are defined by

$$u = \frac{dx}{dt}, \qquad v = \frac{dy}{dt}.$$

If you have not yet covered differentiation then the velocity components can be defined, as in 4.1, as the slopes of the graphs of $x$ and $y$ against time.

Therefore, given $x$ and $y$, $u$ and $v$ can be found and therefore the magnitude (i.e. the speed) and direction of the velocity. As $t$ varies the point with coordinates $x$ and $y$ will describe a curve, the path of the particle, as shown in the diagram.

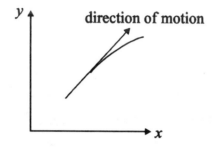

The motion of the particle will be at an angle $\theta$ to the $x$-axis where

$$\tan\theta = \frac{v}{u} = \frac{\frac{dy}{dt}}{\frac{dx}{dt}} = \frac{dy}{dx},$$

therefore the direction of motion of the particle will be along the tangent to the path.

Acceleration is also defined as a vector and its components are $\frac{d^2x}{dt^2}$ and $\frac{d^2y}{dt^2}$. However the acceleration due to gravity is constant and vertically downwards and therefore, if the $x$ and $y$ axes are chosen horizontally and vertically, only one component of acceleration has to be considered in problems of motion under gravity as considered in 8.2.

## Motion under gravity in two dimensions

**Example 8.1**

Find the $x$- and $y$-components of the following velocities
(i) 4 ms$^{-1}$ at an angle of $40°$ to the positive $x$-axis,
(ii) 8 ms$^{-1}$ at an angle of $140°$ to the positive $x$-axis,
(iii) 14 ms$^{-1}$ at an angle of $60°$ below the positive $x$-axis

The velocities are all shown in the diagram below where the $x$-axis is taken to be across the page to the right and the $y$-axis to be up the page.

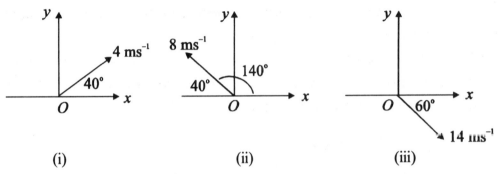

(i)  (ii)  (iii)

(i) The $x$- and $y$-components are
$$4\cos 40° \text{ ms}^{-1} = 3.06 \text{ ms}^{-1} \text{ and } 4\sin 40° \text{ ms}^{-1} = 2.57 \text{ ms}^{-1}.$$

(ii) The $x$- and $y$-components are
$$8\cos 140° \text{ ms}^{-1} = -6.13 \text{ms}^{-1} \text{ and } 8\sin 140° \text{ ms}^{-1} = 5.14 \text{ ms}^{-1}.$$
These components could also have been found, as in problems in Statics, by taking components along the positive $y$ direction and the negative $x$ direction.
These are $\quad 8\cos 40° \text{ ms}^{-1} = 6.13 \text{ ms}^{-1} \text{ and } 8\sin 40° \text{ ms}^{-1} = 5.14 \text{ ms}^{-1}.$
The $x$-component is then found by changing the sign of the first component.

(iii) In this case the simplest way is to find the components along the positive $x$- and negative $y$-axes; these are
$$14\cos 60° \text{ ms}^{-1} = 7 \text{ ms}^{-1} \text{ and } 14\sin 60° \text{ ms}^{-1} = 12.1 \text{ ms}^{-1}$$
The $x$- and $y$-components are therefore 7 ms$^{-1}$ and $-12.1$ ms$^{-1}$.

**Example 8.2**

Find the magnitude and direction of the following velocities given that their $x$- and $y$-components are respectively
(i) 6 ms$^{-1}$, 3 ms$^{-1}$, (ii) $-7$ ms$^{-1}$, 5 ms$^{-1}$, (iii) $-2$ ms$^{-1}$, $-3$ ms$^{-1}$.

## Motion under gravity in two dimensions

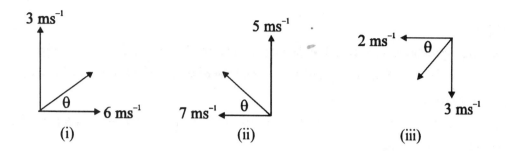

The components are shown in the above diagram where the x-axis is taken to be across the page to the right and the y-axis to be up the page.

(i) The speed is $\sqrt{6^2 + 3^2}$ ms$^{-1}$ = 6.71 ms$^{-1}$ and the motion is at an angle $\theta$ to the positive x-axis where $\tan\theta = \frac{3}{6}$ so that $\theta = 26.6°$ and therefore the velocity is 6.71 ms$^{-1}$ at an angle of 26.6° to the positive x-axis.

(ii) The speed is $\sqrt{7^2 + 5^2}$ ms$^{-1}$ = 8.6 ms$^{-1}$ and the motion is at an angle $\theta$ to the negative x-axis where $\tan\theta = \frac{5}{7}$ so that $\theta = 35.5°$ and therefore the velocity is 8.6 ms$^{-1}$ at an angle of 144.5° to the positive x-axis.

(iii) The speed is $\sqrt{2^2 + 3^2}$ ms$^{-1}$ = 3.61 ms$^{-1}$ and the motion is at an angle $\theta$ to the negative x-axis in the third quadrant where $\tan\theta = \frac{3}{2}$ so that $\theta = 56.3°$ and therefore the velocity is 3.61 ms$^{-1}$ at an angle of 236.3° to the positive x-axis.

### Example 8.3

The x and y displacements, in metres, of a particle from the origin at time $t$ s are $(3t, 4t - 3t^2)$. Find the velocity components when $t = 1$ and $t = 3$. Determine also the direction of motion for $t = 1$.

The velocity components are found by differentiating the displacements and this gives $u = 3$ ms$^{-1}$, $v = (4 - 6t)$ ms$^{-1}$. The velocity components for $t = 1$ are therefore 3 ms$^{-1}$ and $-2$ ms$^{-1}$, and for $t = 3$ are 3 ms$^{-1}$ and $-14$ ms$^{-1}$.

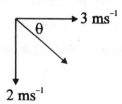

The components for $t = 1$ are shown in the diagram, with the x and y directions as defined in Example 8.2. The velocity is at an angle $\theta$ below the x-axis where $\tan\theta = \frac{2}{3}$ so that $\theta = 33.7°$.

# Exercises 8.1

In the following questions the $x$ and $y$ displacements of a particle from the origin are denoted by $x$ m and $y$ m respectively and the $x$- and $y$-components of velocity by $u$ ms$^{-1}$ and $v$ ms$^{-1}$ respectively

**1** Taking the $x$-axis to be across the page to the right and the $y$-axis to be up the page find $u$ and $v$ in the following cases.

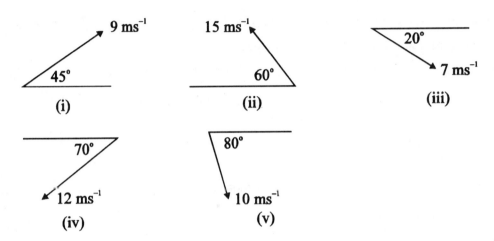

**2** Find the magnitude and directions of the velocities corresponding to
(i) $u = 2$, $v = 6$, (ii) $u = -3$, $v = 8$, (iii) $u = 4$, $v = -12$, (iv) $u = -9$, $v = -4$,
(v) $u = 12$, $v = -15$.

**3** Find $u$ and $v$ when
(i) $x = 5t$, $y = 2t + 4t^2$, (ii) $x = 3t$, $y = 4t - 3t^2$, (iii) $x = 8t^2$, $y = 5t^2 + 4t^3$,
(iv) $x = e^{-t}$, $y = e^{-t} + e^{-2t}$.

**4** Find the speed and the direction of motion for 3(i) and 3(ii) at $t = 2$.

## 8.2 Equations of motion

The motion of any body moving under gravity (such a motion is generally referred to as projectile motion with the moving body being called the projectile) is governed by Newton's second law of motion. You have already met this for motion along a line in the form mass × acceleration. The form for general motion is basically the same except that it is now necessary to take into account that the motion is no longer in a straight line. In more general circumstances Newton's law is

Mass × component of acceleration in any direction = component of force in that direction.

In projectile motion the **modelling assumptions** made are that all bodies are modelled as particles and that there are no resistive forces acting so that the only force acting is that due to gravity. If the $x$-axis is chosen horizontally and the $y$-axis vertically upwards then there is no *x*-component of force and the *y*-component of force is $-mg$, where $g$ is the acceleration due to gravity. Therefore Newton's law gives

mass × *x*-component of acceleration = 0,  mass × *y*-component of acceleration = $-mg$,

i.e.   *x*-component of acceleration = 0,   *y*-component of acceleration = $-g$.

The first equation shows that the *x*-component is a constant e.g. *u*, so that the horizontal displacement of a particle from its initial position is *ut*. The second equation shows that the vertical motion will be the same as that of a particle falling freely under gravity i.e. with downwards acceleration *g*. The vertical displacement can therefore be found from $s = ut + \frac{1}{2}at^2$ with $a = -g$ and *u* equal to the value of the upward velocity component at time $t = 0$; this will be denoted by *v*. Therefore the horizontal and vertical displacements at time *t* of a particle from a point *O* are given by

$$x = ut \quad \text{and} \quad y = vt - \frac{1}{2}gt^2,$$

where *u* and *v* are the components of the velocity of the particle at time $t = 0$ at the initial point *O*. Therefore if a particle were projected from *O* with velocity components *u* and *v* as shown in the left hand diagam its coordinates would be given by the above equations.

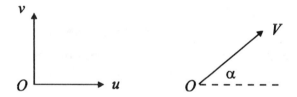

It is often more usual to give, as in the right hand diagram, the magnitude *V* and direction $\alpha$ to the horizontal of the initial velocity. Therefore $u = V \cos \alpha$, $v = V \sin \alpha$ and substituting into the expressions for the dsiplacements gives

$$x = V \cos \alpha\, t \quad \text{and} \quad y = V \sin \alpha\, t - \frac{1}{2}gt^2.$$

Either of the above sets of equations can be used to solve any projectile problem but on the whole it is better to tackle any problem from first principles. It is important to notice that the equations determining the vertical and horizontal displacements are independent of each other. The horizontal motion is a motion at constant speed and the vertical motion is that of a particle moving vertically; you solved problems of this type in 4.3. The actual motion of the particle is the combination of the two independent motions.

It is also possible to generalise the principle of conservation of energy to cover projectile motion and use of this can give a quick method of finding speed. The principle of conservation of energy applies to the vertical motion in the form

$\frac{1}{2}m$ (vertical component of velocity)$^2$ + gravitational P.E = constant.

The horizontal component of velocity is constant and therefore the above equation holds if $\frac{1}{2}m$ (horizontal component of velocity)$^2$ is added to both sides. Also

(horizontal component of velocity)$^2$ + (vertical component of velocity)$^2$ = (speed)$^2$.

The definition of kinetic energy can be generalised to plane motions by

$$\text{K.E.} = \frac{1}{2}m\,(\text{speed})^2,$$

(you will meet the general definition in M2) and therefore with this definition the mechanical energy is conserved.

An alternative but completely equivalent way of finding the displacements would be to use the definition of acceleration in terms of a derivative to get

$$\frac{d^2x}{dt^2} = 0 \quad \text{and} \quad \frac{d^2y}{dt^2} = -g.$$

These can be integrated as in 4.5 to give $x$ and $y$ and obviously this again gives the same expressions as found directly from the constant acceleration formulae.

The expressions found for $x$ and $y$ above can be used to give some general formulae such as that for the greatest height risen. In many examinations full credit will not normally be given for using these formulae without derivation and this is stated in the M1 syllabus. These formulae are however useful in showing the general behaviour of a projectile and their derivation is given in 8.3.

Before working through some examples you may find it useful to have some general idea of the path of a projectile under the assumptions of the model. The actual path will be slightly different mainly due to the effect of air resistance. You will already have some idea of how a projectile moves from watching the motion of a football, tennis ball or cricket ball. You can get a particularly clear picture by looking at the water coming out of hose pipe pointed at an angle.

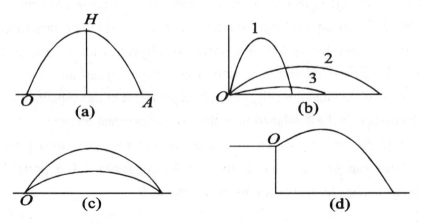

## Motion under gravity in two dimensions

Diagram (a) shows the path taken by a projectile projected from level ground and landing on level ground. It rises to its highest point $H$ and then hits the ground again, the curve described is known as a parabola and it is symmetric about the vertical line through its highest point $H$. At $H$ the particle will be moving horizontally and, in view of the symmetry, the time to reach $H$ will be one half of the time taken to reach $A$. The distance $OA$ from ground to ground is called the range, $R$. Diagram (b) shows the variation in range with different angles of projection for a given speed of projection. Curve 1 shows that for a large angle of projection the range is relatively small. It increases as the angle of projection decreases until it reaches a maximum value for a projection angle of $45°$, this is curve 2. Thereafter the range decreases with the angle of projection as shown by curve 3. Any range other than the maximum value can be attained for a given speed with two angles of projection; if $\alpha°$ is one such angle then $90° - \alpha°$ is the other. This is shown in diagram (c). If the particle is projected from the top of a cliff the path will be as in diagram (d).

The basic method of solving any problem is to write down the displacements from the point of projection at any time. These will involve the initial components of the velocity of projection, or equivalently, the speed and angle of projection. These will either be known or sufficient information will be given to find them. In cases of projection from ground level it is usually better to take the $y$-direction vertically upwards. For problems involving projection from a point above ground level it may be worth taking the $y$-direction downwards. In such a case you should remember to use the correct signs in $s = ut + \frac{1}{2}at^2$.

In all the following examples the horizontal and upwards vertical displacements of a particle at time $t$ s are denoted by $x$ m and $y$ m respectively and $u$ ms$^{-1}$ and $v$ ms$^{-1}$ denote the horizontal and vertical components of the initial velocity and $g$ will be assumed to be 9.8 ms$^{-2}$.


### Example 8.4
A particle is projected from a point $O$ on level ground with velocity components of 7 ms$^{-1}$ horizontally and 19.6 ms$^{-1}$ vertically upwards. Find the distance from $O$ of the point where the particle next hits the ground and also find the greatest height reached above $O$.

Since there is no horizontal component of acceleration the horizontal component of velocity has the constant value 7 and therefore

$$x = 7t.$$

Applying $s = ut + \frac{1}{2}at^2$ with $a = -9.8$, $u = 19.6$ gives
$$y = 19.6t - 4.9t^2.$$
The particle is on the ground when $y = 0$ i.e.
$$19.6t - 4.9t^2 = 0,$$
so that $t = 0$ or $t = 4$, so it next hits the ground when $t = 4$ and $x = 28$. The greatest height is reached at the point where there is no vertical velocity. Applying $v = u + at$ with $u = 19.6$ and $a = -9.8$ shows that the vertical velocity $v$ is given by
$$v = 19.6 - 9.8t.$$
This vanishes when $t = 2$. This result could also have been deduced from the symmetry about the vertical through the point of greatest height. Substituting $t = 2$ into the expression for $y$ gives the greatest height as $(39.2 - 19.6)$ m $= 19.6$ m.
An alternative method for finding the greatest height would have been to use $v^2 = u^2 + 2as$, with $u = 19.6$, $a = -9.8$ and $v = 0$.


## Example 8.5

A particle is projected from a point $O$ with a speed of 25 ms$^{-1}$ at an angle $\theta$ to the horizontal where $\tan \theta = \frac{4}{3}$. Find the equation of the path of the particle and also its direction of motion when $t = 3$.

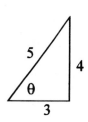
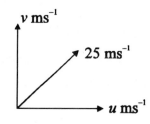

The left hand diagram shows that $\cos \theta = \frac{3}{5}$ and $\sin \theta = \frac{4}{5}$, it then follows from the right hand diagram that the horizontal component of velocity is $25 \cos \theta = 15$ and the vertical component of velocity is $25 \sin \theta = 20$.
The horizontal component of velocity will have the constant value 15 and therefore
$$x = 15\,t.$$
Applying $s = ut + \frac{1}{2}at^2$ with $a = -9.8$, $u = 20$ gives
$$y = 20t - 4.9t^2.$$

## Motion under gravity in two dimensions

Substituting for $t$ in terms of $x$ gives

$$y = \frac{20}{15}x - 4.9\frac{x^2}{225} = \frac{4}{3}x - \frac{4.9x^2}{225},$$

which is the equation of the path.

The vertical component of velocity is found by substituting $u = 20$ and $a = -9.8$ into $v = u + at$ giving

$$20 - 9.8t.$$

(This could also have been obtained by differentiating $y$ with respect to $t$.)

For $t = 3$ the vertical component of velocity is $-9.4$ and the components of the velocity are as shown in the diagram.

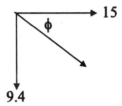

The particle is therefore moving at an angle $\phi$ below the horizontal where $\tan \phi = 0.63$, so that $\phi = 32.1°$.

The direction could also have been obtained by differentiating the equation of the path with respect to $x$ and finding the gradient for $t = 3$ i.e. for $x = 45$.

$$\frac{dy}{dx} = \frac{4}{3} - \frac{9.8x}{225},$$

and its value for $x = 45$ is $-0.63$. The minus sign shows that it is moving below the horizontal.

### Example 8.6

A ball is projected from ground level so that after 2 s it just clears a wall at a distance of 6 m away and 3 m high. Find the initial horizontal and vertical components of velocity.

In this case the initial components of velocity are not given but, as in the previous examples, the coordinates can be expressed in terms of them as

$$x = ut.$$
$$y = vt - 4.9t^2.$$

The conditions give $x = 6$ for $t = 2$, and substituting into the equation for $x$ gives $u = 3$.
Substituting $t = 2$ and $y = 3$ into the equation for $y$ gives

$$3 = 2v - 4.9 \times 4,$$

so that $v = 11.3$.

# Motion under gravity in two dimensions

## Example 8.7

A golf ball is projected from the ground with speed 35 ms$^{-1}$ at an angle $\theta$ to the horizontal where $\tan \theta = \frac{3}{4}$. On its downward path it just clears a tree 5.6 m high. Find the distance of the tree from the point of projection.

The horizontal and vertical components of velocity are given by $u = 35 \cos \theta$ and $v = 35 \sin \theta$. Since $\tan \theta = \frac{3}{4}$, it follows by drawing a right angled triangle as in Example 8.5 that $\cos \theta = \frac{4}{5}$ and $\sin \theta = \frac{3}{5}$ so that $u = 28$ and $v = 21$.

The displacements from the initial point are given by
$$x = 28t.$$
$$y = 21t - 4.9t^2.$$

Substituting $y = 5.6$ into the expression for $y$ gives
$$5.6 = 21t - 4.9t^2.$$

This is a quadratic equation for $t$, this can be solved using the standard formula and the roots are $t = 4$ and $t = 0.29$. The question states that the ball hits the tree on its downwards path so the correct root is the larger one i.e. $t = 4$. Substituting this into the expression for $x$ gives the distance to the foot of the tree as 112 m.

## Example 8.8

A particle is projected horizontally with speed 32 ms$^{-1}$, as shown in the diagram, from the top of a vertical cliff to horizontal ground at a distance of 80 m below the point of projection. Find the distance of the point of impact from the base of the cliff.

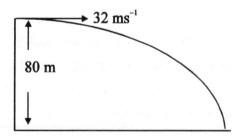

In this case since the motion is entirely downwards, the $y$ direction can be chosen vertically downwards. The initial vertical velocity is zero and therefore the displacements are given by
$$x = 32t,$$
$$y = 4.9t^2.$$

The particle hits the ground when $y = 80$ so that $80 = 4.9t^2$ giving $t = 4.04$. Substituting this in the expression for $x$ shows that the particle hits the ground at a distance of 129.3 m from the base of the cliff.

## Example 8.9

A particle is projected with speed 25 ms$^{-1}$ and with an initial upwards component of velocity of magnitude 19.6 ms$^{-1}$, as shown in the diagram, from the top of a vertical cliff to horizontal ground at a distance of 58.8 m below the point of projection. Find
(i) the distance of the point of impact from the base of the cliff,
(ii) the speed of the particle on impact.

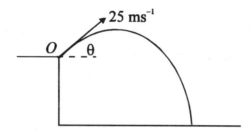

In this question $v$ is given but not $u$. The angle of projection $\theta$ above the horizontal can be found from $19.6 = 25 \sin \theta$ which gives $\theta = 51.6°$ and $u = 25 \cos \theta = 15.52$. In this case the $y$ direction will be taken upwards so that the displacements are

$$x = 15.2t.$$
$$y = 19.6t - 4.9t^2.$$

The particle hits the ground when $y = -58.8$ so that
$$-58.8 = 19.6t - 4.9t^2,$$
this equation simplifies to
$$t^2 - 4t - 12 = 0,$$
which factorises as
$$(t - 6)(t + 2) = 0.$$

The positive root is $t = 6$ and substituting this into the expression for $x$ shows that the particle hits the ground at a distance of 91.2 m from the base of the cliff.

The speed $V$ ms$^{-1}$ on impact can be found using the principle of conservation of energy which gives

$$\tfrac{1}{2} m (25)^2 + m \times 9.8 \times 58.8 = \tfrac{1}{2} mV^2,$$

where $m$ is the mass of the particle and the level of the base of the cliff has been taken as the zero level of potential energy. This equation gives $V = 42.16$.

*Motion under gravity in two dimensions*

## Exercises 8.2

Questions 1 to 6 refer to a particle projected from the origin with horizontal and vertical components of velocity $u$ ms$^{-1}$ and $v$ ms$^{-1}$ (or with speed $V$ ms$^{-1}$ at an angle $\alpha$ above the horizontal), the displacements of the particle at time $t$ s after projection are denoted by $x$ m and $y$ m respectively. Also $g$ should be taken as 9.8 ms$^{-2}$.

**1** $u = 4$, $v = 5$, find $x$ and $y$ in terms of $t$ and find when the particle is next at the level of projection and its horizontal displacement from the point of projection at this time.

**2** $u = 6$, $v = 11$, find $x$ and $y$ in terms of $t$ and find its maximum height above the level of projection.

**3** $u = 4$, $v = 12$, find $x$ and $y$ in terms of $t$ and find the magnitude and direction of its velocity when $t = 1$ and when $t = 8$.

**4** $V = 25$, $\alpha = 30°$, find $x$ and $y$ in terms of $t$ and find when the particle is next at the level of projection and its horizontal displacement from the point of projection at this time.

**5** $V = 30$, $\sin \alpha = \frac{3}{5}$, find $x$ and $y$ in terms of $t$ and find its maximum height above the level of projection.

**6** $V = 40$, $\alpha = 20°$, find $x$ and $y$ in terms of $t$ and find the magnitude and direction of its velocity when $t = 1$ and when $t = 8$.

**7** One second after projection a particle has a horizontal component of velocity of 8 ms$^{-1}$ and an upwards vertical velocity component of 25 ms$^{-1}$. Find the maximum height reached above the point of projection.

**8** A ball is thrown with speed 20 ms$^{-1}$ at an angle of $20°$ above the horizontal and just clears the top of a wall at a distance of 18 m from $O$. Find the height of the wall above the level of $O$.

**9** A stone is thrown from the top of a cliff at a height 75 m above sea level with initial speed 25 ms$^{-1}$ at an angle $\alpha$ above the horizontal where $\cos \alpha = \frac{3}{5}$. Find the distance from the base of the cliff of the point where the stone hits the sea.

**10** A particle projected from the origin and moving under gravity has coordinates (10, 5) two seconds later. Find its initial velocity components.

**11** A particle projected from level ground rises to a height of 19.6 m above it. Find the vertical component of its initial velocity.

**12** A stunt motor cyclist attempts to cross a river 60 m wide by taking off at a speed of 35 ms$^{-1}$ from a ramp at an angle of $25°$ to the horizontal. Determine whether he will be able to cross. He estimates that air resistance will be such that the distance travelled will only be 60% of that predicted by the model neglecting air resistance. Find the minimum speed with which he should leave the ramp.

## 8.3 Basic projectile formulae

In this section some of the basic formulae for projectile motion will be derived and used to establish some of the properties of the path that were described in 8.2. You should be aware of these formulae and know how to derive them but in examinations it is likely that quoting them without proof could bring a penalty.

If a particle is projected from a point $O$ with velocity of magnitude $V$ acting at an angle $\alpha$ above the horizontal then the horizontal and vertical components of its initial velocity are $V \cos \alpha$ and $V \sin \alpha$. The horizontal velocity remains constant and therefore the horizontal displacement is $V \cos \alpha t$.

For the vertical motion applying $s = ut + \frac{1}{2}at^2$ with $a = -g$ and $u = V \sin \alpha$ shows that the vertical displacement is $y = V \sin \alpha t - \frac{1}{2}gt^2$.

Therefore the horizontal and vertical displacements $x$ and $y$ are given by
$$x = V \cos \alpha t, \quad y = V \sin \alpha t - \frac{1}{2}gt^2.$$

The particle will be on the same level as the point of projection at the time $t = T$ when $y = 0$ i.e.
$$V \sin \alpha T - \frac{1}{2}gT^2 = 0.$$

This gives, since the solution $T = 0$ refers to the initial position,
$$T = \frac{2V \sin \alpha}{g},$$

and $T$ is often referred to as the time of flight. Substituting $T$ into the expression for $x$ gives, for a particle projected from ground level, the distance of the point of impact from the point of projection. This is referred to as the range $R$ which is therefore defined by
$$R = \frac{2V^2 \sin \alpha \cos \alpha}{g}.$$

You know from the symmetry properties of the trigonometric functions that
$$\sin\left(\frac{\pi}{2} - \alpha\right) = \cos \alpha \text{ and } \cos\left(\frac{\pi}{2} - \alpha\right) = \sin \alpha,$$

therefore if a particular value of $\alpha$ gives a range $R$ then so will $\frac{\pi}{2} - \alpha$. It will be shown in P2 that $2 \sin \alpha \cos \alpha = \sin 2\alpha$ and therefore the maximum value of $R$, for a given $V$, occurs when $\sin 2\alpha = 1$, i.e. when $\alpha = \frac{\pi}{4}$. A proof, not depending on anything in P2, that the maximum value of range occurs for $\alpha = \frac{\pi}{4}$ is given at the end of this section.

Substituting $u = V \sin \alpha$ and $a = -g$ into $v = u + at$ gives the vertical component of the velocity to be $V \sin \alpha - gt$, this vanishes at the point of maximum height above $O$. Therefore the time to reach this point is $\dfrac{V \sin \alpha}{g}$, which is half the time of flight.

Substituting this value into the expression for $y$ shows that the greatest height $h$ is given by
$$h = \frac{V^2 \sin^2 \alpha}{2g}.$$

(This could also have been obtained by substituting $v = 0$, $u = V \sin \alpha$ and $a = -g$ into $v^2 = u^2 + 2as$.)

Substituting the value of the time to the point of greatest height into the expression for $x$ shows that the displacement of the point of greatest height is $\tfrac{1}{2}R$.

Eliminating $t$ between the expressions for $x$ and $y$ gives
$$y = x \tan \alpha - \frac{gx^2}{2V^2 \cos^2 \alpha} = x \tan \alpha - \frac{gx^2}{2V^2} \sec^2 \alpha,$$
which is the equation of the path of the particle.

You can check, by expanding the squared term in brackets, that the equation of the path can be rewritten as
$$y = -\frac{g \sec^2 \alpha}{2V^2}\left(x - \frac{V^2 \sin \alpha \cos \alpha}{g}\right)^2 + \frac{V^2 \sin^2 \alpha}{2g}$$

This can be further simplified to
$$y - h = -\frac{g \sec^2 \alpha}{2V^2}\left(x - \frac{1}{2}R\right)^2,$$

this shows the symmetry about the point $x = \tfrac{1}{2}R$ i.e. the curve is symmetric about the vertical through the point of greatest height.

**Determination of angle for maximum range**

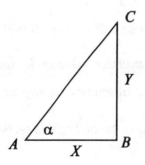

## Motion under gravity in two dimensions

The diagram shows a right angled triangle $ABC$ with the angle $BAC = \alpha$ and $AB = X$ and $BC = Y$. Therefore $\sin \alpha = \dfrac{Y}{\sqrt{X^2 + Y^2}}$, $\cos \alpha = \dfrac{X}{\sqrt{X^2 + Y^2}}$ and

$$2 \sin \alpha \cos \alpha = \dfrac{2XY}{(X^2 + Y^2)}.$$

Also $(X - Y)^2 = X^2 + Y^2 - 2XY$.

The left hand side is positive and attains its minimum value of zero when $X = Y$, therefore

$$\dfrac{2XY}{(X^2 + Y^2)} \leq 1,$$

with the equality holding for $X = Y$. This therefore shows that $2 \sin \alpha \cos \alpha$ has its maximum value when $\cos \alpha = \sin \alpha = \dfrac{1}{\sqrt{2}}$, i.e. for $\alpha = \dfrac{\pi}{4}$.

You may now care to use these results to answer Examples 8.4 and 8.5 and question 1, 2, 5 and 6 of Exercises 8.2

## **Miscellaneous Exercises 8**

**1** A golf ball is struck from a point $O$ on horizontal ground so that initially it is moving with speed 25 ms$^{-1}$ at an angle $\theta$ to the horizontal where $\tan \theta = \dfrac{3}{4}$. Write down expressions for the horizontal and vertical components of its displacement from $O$ at any subsequent time $t$.

State two physical assumptions that you have made in determining the displacements.

Determine the time to reach maximum height and the range as predicted by your model.

**2(a)** A zoo-keeper fires a tranquillising dart at a rhinoceros. The dart is fired at a speed of 25 ms$^{-1}$ at an angle $\alpha$ to the horizontal and, when it hits the rhinoceros, it is at the same height as that at which it was fired. Given that $\cos \alpha = \dfrac{7}{25}$ and $\sin \alpha = \dfrac{24}{25}$ find

 (i) the time of flight of the dart,

 (ii) how close the keeper was to the rhinoceros when the dart was fired.

(The dart is to be modelled as a particle and air resistance may be neglected.)

(b) The zoo-keeper has to tranquillize another rhinoceros well beyond the range of the dart but which is running towards the keeper at a speed of 10 ms$^{-1}$. The keeper fires the dart at the same speed and angle of projection as in (a). How far away should the rhinoceros be from the zoo-keeper when the dart is fired?

**3(a)** At time $t = 0$ a particle $P$ is projected from a point $O$ on horizontal ground with speed 40 ms$^{-1}$ at an angle $\alpha$ to the horizontal where $\cos \alpha = \dfrac{3}{5}$.

## Motion under gravity in two dimensions

Write down expressions for its vertical and horizontal displacements from $O$ at time $t$ seconds and show that it reaches its greatest height after approximately 3.27 s.

Some time after attaining its greatest height the particle hits a screen, which is perpendicular to its plane of motion, at a point $B$ at a height 49.6 m above the horizontal.

(i) Show that the particle hits the screen 4 seconds after being projected.

(ii) Find the gradient of the path immediately before impact.

(b) The screen is then moved nearer to $O$ so that the particle, when projected as above, still strikes it at the point $B$. Find, in metres correct to one decimal place, the perpendicular distance of the screen from $O$ in this case.

(c) The screen is moved to the position where $P$, when projected as above, strikes it at the highest point of its motion. Given that the coefficient of restitution between the particle and the screen is $\frac{1}{4}$ find, in metres correct to one decimal place, the perpendicular distance from the screen to the point where $P$ first hits the ground.

**4** A particle $P$ is projected from a point $O$ on a horizontal plane with speed $v$ at an angle $\alpha$ above the horizontal. It rises to a maximum height $h$ above the plane and strikes the plane again at a distance $r$ from $O$. Write down $x$ and $y$ the horizontal and vertical displacements of $P$ from $O$ at time $t$ after projection, in terms of $v$, $\alpha$, $g$ and $t$. Hence find $r$ and $h$ in terms of $v$, $g$ and $\alpha$.

The particle passes through the point $A$ whose horizontal displacement from $O$ is $pr$, ($0 < p < 0.5$). Find, in terms of $p$ and $\alpha$, the tangent of the angle $\beta$ between the horizontal and the path of the particle at $A$. State the horizontal displacement of the point $B$ where the path is inclined at an angle $\beta$ below the horizontal.

Given that at $A$ the particle is at a height $a$ above the horizontal plane, express $a$ in terms of $p$ and $h$.

**5** A golf ball is driven from a point $O$ with an initial speed of 42 ms$^{-1}$ at an angle $\alpha$ to the horizontal. Neglecting air resistance derive the horizontal component $x$ and the vertical component $y$, of the ball's displacement from $O$ at time $t$ after projection. Show that

$$y = x \tan \alpha - \frac{x^2}{360 \cos^2 \alpha}.$$

The golf ball just clears a tree at $B$ where $B$ is on the same horizontal level as $O$ and $OB$ is 150 m. The tree is 5 m high. Verify that one value of $\alpha$ is such that $\tan \alpha = \frac{3}{5}$.

By using the identity $\frac{1}{\cos^2 \alpha} = 1 + \tan^2 \alpha$ find a quadratic equation satisfied by $\tan \alpha$ and hence find a second value of $\tan \alpha$ such that the ball just clears the tree at $B$.

## Motion under gravity in two dimensions

**6** A particle, projected with speed $V$ at an angle $\alpha$ to the horizontal from $O$, moves freely under gravity. Find $u$ and $v$, the horizontal and vertical components of its velocity, and $x$ and $y$, its horizontal and vertical displacements respectively, at time $t$ after projection. Show that
$$2\frac{y}{x} - \frac{v}{u} = \tan\alpha.$$
Given that the particle strikes a plane through the point of projection and inclined at an angle $\beta$ to the horizontal, where $\tan\beta = \frac{1}{2}$, at right angles find the value of the ratio $\frac{v}{u}$ at the instant of impact. Hence find the value of $\tan\alpha$.

**7** A particle $P$ is projected with speed $V$ at an angle $\alpha$ to the horizontal from $O$. Find $u$ and $v$ the horizontal and vertical components of its velocity, and $x$ and $y$, its horizontal and vertical displacements respectively, at time $t$ after projection.
Given that the particle strikes the horizontal plane through $O$ after a time $T$ show that
$$T = \frac{2V\sin\alpha}{g}.$$
Find, in terms of $g$ and $T$, the maximum height of the particle above the level of $O$.
Given that at time $\frac{5T}{8}$ the particle is moving at right angles to its original direction of motion find $\tan\alpha$.

**8** A particle is projected from a point $O$ on level ground and next strikes the ground again after a time $T$ at the point $A$, where $OA = 2a$. Find the horizontal and vertical components of the velocity of projection.

**9** A stone is projected with a velocity of $14.7$ ms$^{-1}$ at an angle $\alpha$ to the horizontal where $\sin\alpha = \frac{3}{5}$. Before reaching its maximum height it just misses the top of a pole of height 10 m. At the instant the stone is thrown a bird leaves the top of the pole and flies horizontally at a constant speed of $v$ ms$^{-1}$. Find $v$ given that the stone hits the bird.

**10** Two seconds after a stone is thrown it is moving at an angle $\alpha$ to the horizontal where $\tan\alpha = 2$, a further second later it is moving at an angle $\beta$ to the horizontal, where $\tan\beta = 1$.
Find, by considering the ratio $\frac{\tan\alpha}{\tan\beta}$, the vertical component of the initial velocity of projection. Determine also the horizontal component of the initial velocity of projection.

**11** A particle is projected from a point with speed 25 ms$^{-1}$ at an angle $\alpha$ to the horizontal where $\tan\alpha = \frac{7}{25}$. Find the magnitude and direction of the velocity of the particle two seconds after projection.

**12** At a particular instant a particle $P$ is projected from a point $O$ with horizontal and upward components of velocity $3nu$ and $5nu$ respectively, where $n$ and $u$ are positive constants. At the same instant a second particle $Q$ is projected from a point whose coordinates referred to $O$ are $(16a, 17a)$, where the $x$-axis is horizontal and the $y$-axis is vertically upwards.

The initial $x$ and $y$ components of the velocity of $Q$ are $-4u$ and $3u$ respectively.

Find the horizontal and vertical components of the displacements from $O$ of $P$ and $Q$ at time $t$ after projection and find the value of $n$ such that the particles collide.

**13(a)** A particle is projected from a point $O$ with speed $u$ at an angle $\alpha$ above the horizontal.

(i) Write down expressions for the horizontal and vertical displacements of the particle from $O$ at time $t$ after projection,

(ii) deduce that the particle first hits the horizontal plane through $O$ at a distance
$$\frac{2u^2 \sin \alpha \cos \alpha}{g}$$
from $O$,

(iii) show that the greatest height reached by the particle above the level of $O$ is
$$\frac{u^2 \sin^2 \alpha}{2g}.$$

(b) A particle $P$ is projected from a point $O$ with speed $\sqrt{12ga}$ at the angle which gives maximum range on the horizontal plane through $O$. Find the tangent of the angle between the velocity of $P$ and the horizontal at time $\sqrt{\dfrac{3a}{8g}}$ after projection.

(c) A particle $Q$ is projected with speed $\sqrt{12ga}$ from a point $A$ on the horizontal floor of a room with a horizontal ceiling at a height $\dfrac{3a}{2}$ above the floor. Find, assuming that $Q$ must not hit the ceiling, the maximum value of the distance from $A$ of the point at which $Q$ first hits the floor.

**14**

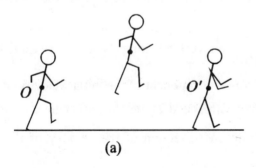

(a)

Diagram (a) shows a simplified schematic diagram of a model of the action of a long jumper. The points $O$ and $O'$ denote the positions of the centre of gravity of the jumper at the start and the end of the jump respectively. In this model it is assumed that the points $O$ and $O'$ are on the same horizontal level and that the only force acting on the jumper during the jump is the force due to gravity. The jumper is taken to be a particle occupying the position of his centre of gravity and projected at time $t = 0$ s with speed $V$ at an angle $\alpha$ to the horizontal. The horizontal and vertical components of the displacement from $O$ at time $t$ are denoted by $x$ and $y$ respectively. Write down expressions for $x$ and $y$ at time $t$ and hence show that
$$y = px - qx^2,$$
where $p$ and $q$ are constants which should be found in terms of $V$, $g$ and $\alpha$.

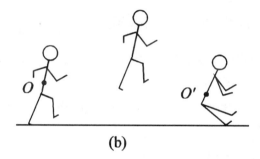

(b)

Diagram (b) shows a more realistic model where it is assumed that the centre of gravity of the jumper drops a vertical distance of 0.3 m between take-off and landing. In one particular instance the values of $\alpha$ and $V$ (which you do not need to find) are such that the above equation, with $x$ and $y$ measured in metres, becomes
$$y = \frac{4}{3}x - \frac{x^2}{45}.$$
Find the difference between the horizontal displacement of the centre of gravity between take-off and landing calculated using this second model and that calculated using the first model.

**15** A golf ball is at rest at a point $A$ on horizontal ground. Some distance away is a tree, 17.5 m high. The golf ball is struck so that the horizontal and vertical components of its initial velocity are 24.5 ms$^{-1}$ and 28 ms$^{-1}$ respectively. The golf ball just clears the top of the tree when it is on the downward part of its flight. Find

(a) the time taken by the ball to reach the top of the tree,
(b) the distance of the base of the tree from $A$,
(c) the distance of the base of the tree from the point where the ball strikes the ground.

**16** A particle $P$ is projected with speed 49 ms$^{-1}$ at an angle $\alpha$ to the horizontal from $O$. Find $x$ and $y$, its horizontal and vertical displacements respectively, at time $t$ s after projection. Hence show that for $x = 140$
$$y = 140 \tan \alpha - 40 \sec^2 \alpha.$$
The particle just clears a wall 20 m high at a distance of 140 m from the point of projection. Find, by using the result $\sec^2 \alpha = 1 + \tan^2 \alpha$, the two values of $\tan \alpha$ for which this is possible.

**17** A point $O$ is vertically above a fixed point $A$ of a horizontal plane and a particle $P$ is projected from $O$ with speed $5V$ at an angle $\alpha$ above the horizontal where $\cos \alpha = \frac{3}{5}$. It hits the plane at a point $B$ at a distance $\frac{48V^2}{g}$ from $A$. Show that the height of $O$ above $A$ is $\frac{64V^2}{g}$ and find the distance of $P$ from $O$ when it is directly level with it.

A second particle is now projected with speed $24W$ from $O$ at an angle $\alpha$ above the horizontal and it also hits the plane at $B$. Find an equation involving $V$, $W$, $g$ and $\alpha$. Given that one value of $\alpha$ is $45°$ find $V$ in terms of $W$ and show that the other value of $\alpha$ is such that
$$7 \tan^2 \alpha - 6 \tan \alpha - 1 = 0.$$

**18** A particle is projected from a point $O$ on a horizontal plane with speed $u$ in a direction making an angle $\alpha$ above the horizontal. At a subsequent time $t$ the horizontal and vertical displacements of the particle from $O$ are denoted by $x$ and $y$ respectively and it is moving in a direction inclined at an angle $\beta$ above the horizontal, with the upward vertical component of its velocity being $v$.

(a) Show, from the equations of motion, that

(i) $v + u \sin \alpha = \dfrac{2y}{t}$,

(ii) $x(\tan \alpha + \tan \beta) = 2y$.

(b) When moving at an angle of $45°$ above the horizontal the particle just clears a wall of height 3 m and at a perpendicular distance of 2 m from $O$. The wall is perpendicular to the plane of motion of the particle. Subsequently, when moving at an angle of $45°$ below the horizontal the particle just clears a second identical and parallel wall. Find

   (i) $\tan \alpha$,
   (ii) the distance between the walls,
   (iii) the range on the horizontal plane through $O$,
   (iv) the maximum height reached above the plane.

# INDEX

| | |
|---|---|
| Acceleration | 73, 184 |
|     constant | 74 |
|     due to gravity | 43, 82 |
| Collisions | 165, 172 |
| Connected bodies | 104, 167 |
| Conservation of energy | 139, 189 |
| Conservation of momentum | 163, 164 |
| Displacement | 70 |
| Elastic energy | 138 |
| Energy | |
|     conservation | 139, 189 |
|     gravitational/potential | 138 |
|     kinetic | 132 |
| Equilibrium | 18, 62 |
| Force | 7 |
|     components | 12 |
|     moment | 56 |
|     of gravity | 24, 43 |
|     resultant | 10, 39 |
| Friction | 26, 48 |
|     coefficient | 26, 49 |
|     limiting | 26, 50 |
| Hooke's Law | 9, 25, 46 |
| Impulse | 158, 159 |
| Impulse momentum principle | 158 |
| Kinetic energy | 132 |
| Light | 24 |
| Mass | 96, 116 |
| Modulus of elasticity | 25 |
| Momentum | 158 |
| Newton, unit of force | 9, 97 |
| Newton's laws of motion | 25, 96, 97 |
|     elastic law | 172, 173 |
| Particle model | 3 |
| Peg, smooth | 25, 48 |
| Potential energy | 138 |
|     gravitational | 138 |
|     elastic | 138 |
| Power | 111, 149 |
| Projectile | 187 |
| Pulley, small | 108 |
|     smooth | 25, 48 |
| Restitution, coefficient of | 172, 173 |
| Rod, light | 25, 47 |
| Supports, simple | 25 |
| Spring | 25 |
| String | 24 |
|     light | 24 |
|     inextensible | 24 |
|     elastic | 25, 45 |
| Surfaces, smooth | 25 |
| Tension | 24 |

Velocity 72, 181
Velocity time diagram 75

Weight 43
Work done
    by a constant force 125
    by a variable force 129
    stretching a string 131
Work energy principle 133, 139

# Answers to Exercises

### Exercises 2.1
1. 2.85 N, 10.63 N   2 −4 N, −6.93 N   3 −4.79 N, 13.16 N
4  5.13 N, −14.10 N   5 (a) −5 N, −8.66 N  (b) −4.60 N, −3.86 N
6 (a) 4 N, −6.93 N  (b) 1.37 N, −3.76 N

### Exercises 2.2
1  7 N, 2 N    2  14 N, 6.93 N   3  1.41 N, −1.01 N
4  1.73 N, −2 N   5  −8.57 N, −4.55 N

### Exercises 2.3
1  12 N    2  $\alpha = 0, P = 6$ N   3  $P = 11.28$ N, $Q = 4.1$ N
4  $P = 17.07$ N, $Q = 12.78$ N    5  5.96 N, $Q = 4.39$ N
6  26 N, 67.4° N   7  14.74 N, 4°

### Exercises 2.4
1 (a) 3.92 N, 3.92 N  (b) 14.7 N   2 (a) 29.4 N  (b) 2 kg
3  0.4 m    4  1000 N   5  1.4 m   6  3.29 N
7  $P = 1.96$ N, $T = 3.39$ N   8  2.78 N, 3.08 N   9  $\dfrac{mg}{(1+\cos\alpha)}\sin\alpha$
10  2.08 m   11  2.09 m   12  509.2 N, horizontally; 882 N vertically
13  261.3 N   14  3.53 kg   15  0.2   16  0.11

### Exercises 2.5
1 (a) 30.64 N  (b) 78.32 N   2  0.34    3 (i) 7.84 N  (ii) 9.22 N
4 (i) 17.79 N  (ii) 27.25 N   5 (i) 100.3 N  (ii) 11.49 N   6  0.51

### Exercises 2.6
1 (a) 7.62 N, 23.2°  (b) 8.94 N, −63.4°  (c) 11.4 N, 105°  (d) 13.9 N, 249°
   (e) 8.06 N, 150.3°  (f) 4.24 N, 225°
2 (a) 3.73 N, −77.8°  (b) 5.95 N, 141°   (c) 5.91 N, 72.5°
3 (a) 3.73 N at 12.2°  (b) 5.95 N at −39°  (c) 5.91 N at 252.5°

### Miscellaneous Exercises 2
1. (a) 20 N, 21 N  (b) 29 N  (c) 46.4°  (d) 29 N at 226.4° to $x$ direction
2  1, 7    3  42 N at 261.8° to $AB$    4  22, 60°
5  light, 0.48 m    6  78 N, 325 N   7 (i) $\sqrt{3}P$  (ii) 150°  (iii) 120°, 60°
9  16.7°    10  4022 N, 1105 N   11  31°    12  0.24    13  0.58
14  11.4 N, 13.9 N    15 (a) 80 N  (b) $20\sqrt{3}$ N   16  0.5    17  $\dfrac{1}{4}, \dfrac{3}{\sqrt{15}}$
18 (i) 1975 N  (ii) 3890 N  (iii) 3712 N   20 (i) 19.6 N  (ii) 41.8°
21 (i) 4.62 N, −63 N    22  1.73 kN

## Exercises 3.1
1  − 6.4 Nm, − 2.4 Nm    2  3.8 Nm, − 7.6 Nm    3  1.6 Nm, 9.6 Nm
4  − 3.3 Nm, − 3.9 Nm    5  5.2 Nm    6  26 Nm    7  − 3 Nm
8  18 Nm    9  11 Nm    10  − 1 Nm    11  $2Ql\cos\theta - Wl\cos\theta$, $Ql\cos\theta - Pl\sin\theta$
12  $2Ql\cos\theta - Wl\cos\theta - 2Fl\sin\theta$, $Ql\cos\theta - Pl\sin\theta - Fl\sin\theta$
13  $4Sl\cos\theta - 10Wl\cos\theta$, $6Wl\cos\theta - 4Rl\cos\theta$
14  $2Ra - Wa\sin 2\theta$, $2Fa\sin\theta - Wa\sin 2\theta$

## Exercises 3.2
1  10 N, 6 N    2  13 N, 4 N    3  1 N, 15 N    4  6 N, 3 N
5  4 kg    6  $2\tfrac{2}{3}$ m from $B$    7  $1\tfrac{2}{3}$ m
8  21.82 N, 38.18 N, 40 N    9  $m\tan 15°$

## Exercises 3.3
1  12 N at 3.5 m to right of $O$    2  16 N downwards at 3.5 m to left of $O$
3  26 N upwards at 8 m to right of $O$    4  12 N downwards at 4.5 m left of $O$
5  4 N upwards at 11 m to right of $O$

## Miscellaneous Exercises 3
1  128.7 N, 91.92 N    2  240 N, 460 N    3  64 N, 220 N, 300 N
4  4 m from $A$    5  4.35 kNm    6  1421 N, 1274 N
7  1274 N, 686 N    8  $2W, 6W, \dfrac{4W}{3}, \dfrac{16W}{3}$
9  (a) 7.5 kN, 10.5 kN  (b) 7.5 kN    10  75 tonne, 3.75 m
11  (a) 105 N  (b) 2.91 m  (c) 105 N, 210 N    12  5 kg, $\dfrac{12a}{5}$
13  $R = 1$ N, $S = 3.5$ N, $P = 4.5$ N    14  18.4°    15  $\dfrac{7Wa}{4a-x}, \dfrac{Wa}{2a-x}$

## Exercises 4.1
1  2    2  3, − 13.5    3  14    4  27.5    5  2.5    6  2
7  2    8  $\dfrac{10}{3}$    9  4    10  1.6, 2
11  $u = 12, a = 0.5, u = 13, a = 0$    12  $t = 6$    13  $t = 3, 6$
14  $\dfrac{20}{3}$ ms$^{-2}$, 4ms$^{-2}$, 1800 m    15  32 s, 12 s, 16 s    16  5s
17  15 ms$^{-1}$, 0.15 ms$^{-2}$    18  10 ms$^{-1}$, 300 s    19  21 s

**Exercises 4.2**

1. 20.4 m, 4.1 s
2. 29.7, 2 s
3. (a) 1 s, 8 s  (b) 6.2 s, 2.9 s
4. 39.4
5. 6.1 s
6. 14.9 ms$^{-1}$, 1 s
7. 433 m
8. $\dfrac{u}{g} + \dfrac{1}{2}T$
9. 2.3 m
10. 22.5 m
11. 2 s, 4.4 m from base
12. 7.35 m

**Exercises 4.3**

1. $84t^2 + 12t$
2. $9t^2 + 8t$
3. $2t^4 + 3t^3 + t^2 + 2t$
4. $t^5 + t^4 - 3t + 1$
5. $1 + 5t + e^{-t}$
6. 2, 9
7. $5t^3 + 2t + 2$

**Miscellaneous Exercises 4**

1. (a) 1.1 ms$^{-1}$  (b) 0.95 m  (c) $8.2 - 1.2t - 0.2t^2$  (d) 3, 5
2. $\dfrac{47}{3}, \dfrac{40}{3}$ m
3. 905.6 m
4. 126 m
5. (i) 16  (ii) 0.8 ms$^{-2}$
6. (i) 20 ms$^{-1}$  (ii) 0.25 ms$^{-2}$, 2200 m
7. $\dfrac{192}{u+6}$ s, $\dfrac{60}{u}$ s  (i) 10  (ii) $\dfrac{1}{3}$ ms$^{-2}$, $\dfrac{5}{3}$ ms$^{-2}$
9. $\dfrac{10}{u}, \dfrac{14}{u}, \dfrac{6}{u}$, 5, $\dfrac{5}{2}$ ms$^{-2}$, $\dfrac{25}{6}$ ms$^{-2}$
10. (i) $\dfrac{1}{2}ft^2 + b - ut$  (ii) $\dfrac{v^2}{2f} + (v-u)t + b$, $b + \dfrac{u^2}{2f}$
11. 22.5 s
12. (b) 0.4 ms$^{-2}$, 1125 m  (c) 825 s  (d) 960 s
13. 104.94 m
14. 6, $\dfrac{2}{3}$
15. (i) 11.025 m  (ii) 1.5 s  (iii) 4.9 ms$^{-1}$
16. (i) 0.5 m  (ii) 1 s
17. $\dfrac{351U^2}{800g}, \dfrac{5U}{4}$
18. $19.6t - 4.9t^2$, $19.6(t-2) - 4.9(t-2)^2$, 3, 9.8 ms$^{-1}$
19. 8 cm
20. (a) 2 m  (b) 14 ms$^{-2}$
21. $\dfrac{32}{27}$, 2, 4 ms$^{-1}$
22. $4 \ln \dfrac{4}{3}$
24. $2 < t < 5$, 8.4
25. $p = 4, q = -\dfrac{8}{3}, r = \dfrac{4}{9}$

**Exercises 5.1**

1. 1800 N
2. 2 ms$^{-2}$
3. (i) 45  (ii) 40 m
4. 134 N
5. (i) 3 ms$^{-2}$  (ii) 1300 kg  (iii) 96 m
6. (i) 313.6 N  (ii) 313.6 N
7. 118 N, 98 N, 95.5 N
8. 1404 N, 1420 N, 1388 N
9. 3
10. 0.057
11. 10.2 ms$^{-2}$
12. $g$
13. 18.375 m, 0.23
14. 2.12 ms$^{-2}$
15. At 53.1° to force of magnitude 30 N and of magnitude 100 ms$^{-2}$.

**Exercises 5.2**

1  $0.25$ ms$^{-2}$   2 (i) 1260 N   (ii) 3180 N
3 (i) 1560 N   (ii) 520 N   4  1000 N, 500 N, 700 N
5 (a)(i) 2400 N   (ii) 1600 N   (iii) 800 N   (b)(i) 17400 N   (ii) 11600 N   (iii) 5800 N
6  $0.9$ ms$^{-2}$, 385 N

**Exercises 5.3**

1  $2.45$ ms$^{-2}$, 36.75 N   2  $3.92$ ms$^{-2}$, 41.2 N
3  $0.2g$, $4.8Mg$   4  $6.1$ ms$^{-2}$, 18.4 N   5  11.1 N
6  29.4 N   7  $6.9$ ms$^{-2}$   8  33.1 N   9  $6.3$ ms$^{-2}$   10  $2.45$ ms$^{-2}$

**Exercises 5.4**

1  9 kW   2  9 kW   3  500 N   4  8 ms$^{-1}$   5  $0.2$ ms$^{-2}$
6  12.5 kW, $9.3$ ms$^{-1}$   7  48 kN, 480 kW, 20 ms$^{-1}$
8  $20.5$ ms$^{-1}$, 0.24   9  $32.1$ ms$^{-1}$, $0.96$ ms$^{-2}$   10  50 kW, $0.02$ ms$^{-2}$, 0.036
11  18.24 kW   12  $0.1$ ms$^{-2}$   13  20 ms$^{-1}$

**Exercises 5.5**

1  4 ms$^{-1}$   2  9 ms$^{-1}$, 0   3  $40.24$ ms$^{-1}$, 2.48 m
4 (i) 6 ms$^{-1}$   (ii) $6t - 1.5t^2$   (iii) 8 m


**Miscellaneous Exercises 5**

1  4900 N, 13, 300 N   2  720, 50 s   3  $\frac{10}{3}$ s, $\frac{130}{3}$ m
4  $\frac{1}{8}$ ms$^{-2}$, $\frac{1}{8}$ ms$^{-2}$, 198.4 N, 196 N, 193.5 N
5  $0.65$ ms$^{-2}$, $0.65$ ms$^{-2}$,   6  1.6 kg, $0.89$ ms$^{-2}$
7  $v^2 > 2a\left(\frac{R}{m} + g\right)$   8 (a) 470.4 N   (b) 494.9 N, 382 N
9 (a)(i) 1 ms$^{-2}$   (ii) 1250 m   (b)(i) $\frac{5}{3}$ ms$^{-2}$   (ii) 12 kN
10  0.37   11  $\frac{P}{20}$, 14000, $0.6$ ms$^{-2}$   12  $100(700 + R)$, 300
13  $\frac{1000R}{v} - Mg \sin \alpha - Ma$ newtons   14 (i) 20 ms$^{-1}$   (ii) $0.75$ ms$^{-2}$
15  38 kW   (i) $0.475$ ms$^{-2}$   (ii) 1382.5 N, No, $m \geq 2000$
16  20 kW, $19.44$ ms$^{-1}$, 800 N, $0.39$ ms$^{-2}$
17  500 N, 36.9 kW   18  35 kW, 0.69°, $-0.59$
19  300 N, (i) $0.2$ ms$^{-2}$   (ii) 255 N   20  $\frac{3g}{5}$, $\frac{16mg}{5}$

**21** $N = 7$, $T = 38016$    **22** (a)(i) light  (ii) smooth, $\dfrac{3g}{5}$

**24** $6mg$, $5a$    **25** (a) $\dfrac{2g}{3}$, $\dfrac{2mg}{3}$  (b) $\dfrac{4g}{9}$, $\dfrac{7mg}{9}$

**26** $\dfrac{3g}{10}$, $2.6\,mg$, $\dfrac{2d}{5}$    **27** $\dfrac{128}{3}$ m, $0.4t$ N

**28** (a) $5\,\text{ms}^{-1}$, $20\,\text{ms}^{-1}$  (b) $2\,\text{m}$, $9\,\text{m}$, $11\,\text{m}$

**29** $\left(\dfrac{P-R}{M}\right)\dfrac{T}{2}$, $\left(\dfrac{P-R}{M}\right)\dfrac{T^2}{3}$, $\dfrac{3\left(P^2-R^2\right)T}{16M}$, $\dfrac{R(P-R)T}{2M}$

**30** 13.3 kW

### Exercises 6.1
**1** 1.6 J    **2** 1.86 m    **3** 245 J    **4** 11.76 J    **5** 3.02 J
**6** 26.1 N    **7** (i) 804.4 J  (ii) 1326 J  (a) 177.5 N  (b) 251 N
**8** 424.7 J    **9** 69 kJ    **10** 9.125 N

### Exercises 6.2
**1** (i) 6 J  (ii) −14 J    **2** (i) 22 J  (ii) 5.5 J  (iii) 1 J
**3** 550 J

### Exercises 6.3
**1** 1 J    **2** 8 J    **3** 39.1 J    **4** 9 J
**5** 33 J    **6** −15 J, 2.25 J    **7** 0.25 J

### Exercises 6.4
**1** (i) 6.4 J  (ii) 193.6 kJ  (iii) 440 J    **2** 67.5 kJ    **3** $18.7\,\text{ms}^{-1}$
**4** (i) 2 kN  (ii) $80000x$ where $x$ m is distance entered.    **5** 68400 J
**6** (i) 1132 J  (ii) $1.5\,\text{ms}^{-1}$    **7** $9.9\,\text{ms}^{-1}$
**8** 73.2 J, $2.05\,\text{ms}^{-1}$    **9** 2.36 J, $8.74\,\text{ms}^{-1}$    **10** (i) 2.48 J  (ii) 0.28
**11** 13.2 m    **12** $9.23\,\text{ms}^{-1}$    **13** $u > 4.47$

### Exercises 6.5
**1** 4.41 J    **2** −20.79 kJ    **3** $14\,\text{ms}^{-1}$    **4** 1.98 m
**5** 1.46 m    **6** $\sqrt{gh}$    **7** $0.75\,\text{ms}^{-1}$    **8** $10\,\text{ms}^{-1}$
**9** $.85(ga)^{1/2}$    **10** 2.73 m    **11** 3m    **12** 1.73 m
**13** 3 m    **14** 1.805 m

### Exercises 6.6
**1** $24\exp(6t)$    **2** (i) 30 kJ  (ii) 48.96 kJ    **3** $50.8\,\text{ms}^{-1}$
**4** (i) 25.43 kW  (ii) 33.9 kW    **5** 9.43 kW

## Miscellaneous Exercises 6

1 (a) 1.28 m  (b) 8.01 ms$^{-1}$   2 (b) 5.1 m  (c) 1.18 J
3 1860 N, 93 N    4 (a) 59150 J  (b) 172323,  93.3 N
5 64 kJ, 29.4 J, 126.6 kJ   6 15 J   7 $\dfrac{mgR}{6}$, $\sqrt{\dfrac{gR}{3}}$
8 8 ms$^{-1}$    9 313.6 N, 0.245 m    10 (i) 3.16 ms$^{-1}$  (ii) 2.9 ms$^{-1}$
11 $-3.19$ ms$^{-2}$, 3.64 ms$^{-1}$   12 2a   14 4.5$l$
15 (i) $\sqrt{2gx}$   16 $\sqrt{\dfrac{18ga}{5}}$   17 (i) $\dfrac{7a}{9}$ below $O$  (ii) $\dfrac{2}{3}(ag)^{1/2}$
18 40500 J, 7840 J, 48.34 kW   19 0.14 J

## Exercises 7.1

1 (a) 0.12 Ns  (b) 7.5 Ns  (c) 30 000 Ns   2 1 Ns    3 $-9.6$ Ns
4 $-4.4$ Ns   5 1.6 Ns   6 4 ms$^{-1}$   7 10 500 Ns
8 2.86 ms$^{-1}$, 11.45 m   9 1.44 Ns   10 (i) 36 N  (ii) $c = 1800$
11 5.7 N   12 $k = 2.74 \times 10^5$   13 3.4 Ns   14 2.1 Ns

## Exercises 7.2

1 2    2 1.25    3 3.71    4 3.29, 1.54 J
5 2.2, 29.4 J   6 2.6, 72.6 J   7 8.3 ms$^{-1}$
8 0.86 ms$^{-1}$   9 2.3 kmh$^{-1}$   10 3.79 ms$^{-1}$   11 0.038

## Exercises 7.3

1 4.57 ms$^{-1}$, 13.7 Ns    2 2.09 ms$^{-1}$, 0.42 Ns
3 0.32 m   4 1.01 s, 1.86 ms$^{-1}$    5 2.86 s    6 0.3 m

## Exercises 7.4

1 0.8 m   2 0.79   3 1.13 m   4 1.92 Ns   5 4.2 ms$^{-1}$
6 4.25, 6.25   7 0.25, 4   8 4.29, 0.29   9 1.8, 3.8
10 0.47, 3.6   11 9, 0.17   12 $e > \dfrac{1}{2}$
13 0.11 ms$^{-1}$, 0.12 ms$^{-1}$, 0.77 ms$^{-1}$

## Miscellaneous Exercises 7

1 9.9 ms$^{-1}$, 21.9 ms$^{-1}$   2 1.15 ms$^{-1}$, 1385 Ns
3 23.2 ms$^{-1}$, 2320 Ns   4 (i) 3 ms$^{-1}$  (ii) 36000 Ns  (iii) 120 kN
5 $\dfrac{g}{9}$, $40\dfrac{mg}{9}$, $\dfrac{9V}{11}$   6 $\dfrac{3g}{10}$, 2.6$mg$, $\dfrac{2d}{5}$, $\dfrac{2}{5}\sqrt{\dfrac{gd}{5}}$   7 3$u$
8 (a) 3.4 ms$^{-1}$  (b) 0.5   9 $\dfrac{1}{2}mgh$, $3(1+\sqrt{2})\sqrt{gh}$, $\dfrac{1}{2}\sqrt{gh}$

**10** 0.71    **11** $\frac{20u}{3}$, $\frac{2}{3}$    **13** 0.44 m

**14** (a) $u|4a-2|$, $u|2a-3|$  (b) $\frac{6a}{5}-1$  (c) $\frac{5}{6} \le a \le \frac{5}{3}$  (d) 1

**15** $\frac{5u}{2}$, $\frac{7}{8}$, $\frac{3}{2}mu^2$    **16** $q > 20$

**17** $8mu$, 3.33, 0.19    **18** (a) $2.8u$  (b) $6mu$  (c) 0.91

**19** $\frac{1.2u}{k+1}$  (i)(a) 2 m  (b) 0.75 m  (c) 6.2    **20** (i) 15 ms$^{-1}$  (ii) 3.2 Ns

### Exercises 8.1

**1** (i) 6.36, 6.36  (ii) −7.5, 13  (iii) 6.58, −2.39  (iv) −4.1, −11.28  (v) 1.74, −9.85
**2** (i) 6.32 ms$^{-1}$ at 71.6° to $Ox$  (ii) 8.54 ms$^{-1}$ at 110° to $Ox$  (iii) 12.6 ms$^{-1}$ at 71.6° below $Ox$  (iv) 9.85 ms$^{-1}$ at 204° to $Ox$  (v) 19.21 ms$^{-1}$ at 51.3° below $Ox$
**3** (i) 5, $2 + 8t$  (ii) 3, $4 - 6t$  (iii) $16t$, $10t + 12t^2$  (iv) $-e^{-t}$, $-e^{-t} - 2e^{-2t}$
**4** 18.68 ms$^{-1}$ at 74.5° to $Ox$, 8.54 ms$^{-1}$ at 69.4° below $Ox$

### Exercises 8.2

**1** $4t$, $5t - 4.9t^2$, $t = 1.02$, 4.8 m      **2** $6t$, $11t - 4.9t^2$, 6.21 m
**3** $4t$, $12t - 4.9t^2$, 8.2 ms$^{-1}$ at 41.6° above horizontal, 218.6 ms$^{-1}$ at 89° below horizontal
**4** $21.65t$, $12.5t - 4.9t^2$, $t = 2.61$, 55.2 m
**5** $24t$, $18t - 4.9t^2$, $t = 3.7$, 16.5 m
**6** $37.59t$, $13.68t - 4.9t^2$, 38.6 ms$^{-1}$ at 13.1° above horizontanl, 207.6 ms$^{-1}$ at 79.6° below horizontal
**7** 61.8 m    **8** 2.04 m    **9** 602.5 m    **10** 5 ms$^{-1}$, 12.3 ms$^{-1}$
**11** 19.6 ms$^{-1}$    **12** Yes, 35.8 ms$^{-1}$

### Miscellaneous Exercises 8

**1** $20t$, $15t - 4.9t^2$, 1.53 s, 61.2 m    **2** (a)(i) 4.9 s, 34.3 m  (b) 83.3 m
**3** (a) $24t$, $32t - 4.9t^2$, (ii) −3  (b) 60.7 m  (c) 19.6 m
**4** As in 8.3, −2, 3      **7** 2      **8** $\frac{2a}{T}$, $\frac{gT}{2}$      **9** 9.7 ms$^{-1}$
**10** $4g$, $g$    **11** 27.1 ms$^{-1}$ at 27.7° below horizontal
**12** $3mut$, $5mut - \frac{1}{2}gt^2$, $16a - 4ut$, $17a + 3ut - \frac{1}{2}gt^2$, 4
**13** (a) As in 8.3  (b) $\frac{3}{4}$  (c) $10.4a$    **14** (a) $\tan \alpha$, $\frac{g}{2V^2 \cos^2 \alpha}$  (b) 0.2 m
**15** (a) 5 s  (b) 225 m  (c) 174 m    **16** 0.5, 3    **17** $W\sqrt{28}$
**18** (b)(i) 2  (ii) 4 m  (iii) 8 m  (iv) 4 m